SUSTAINABILITY

Utilizing Lean Six Sigma Techniques

Industrial Innovation Series

Series Editor

Adedeji B. Badiru

Department of Systems and Engineering Management
Air Force Institute of Technology (AFIT) – Dayton, Ohio

PUBLISHED TITLES

Carbon Footprint Analysis: Concepts, Methods, Implementation, and Case Studies, *Matthew John Franchetti*

Computational Economic Analysis for Engineering and Industry, *Adedeji B. Badiru & Olufemi A. Omitaomu*

Conveyors: Applications, Selection, and Integration, *Patrick M. McGuire*

Global Engineering: Design, Decision Making, and Communication, *Carlos Acosta, V. Jorge Leon, Charles Conrad, and Cesar O. Malave*

Handbook of Industrial Engineering Equations, Formulas, and Calculations, *Adedeji B. Badiru & Olufemi A. Omitaomu*

Handbook of Industrial and Systems Engineering, *Adedeji B. Badiru*

Handbook of Military Industrial Engineering, *Adedeji B.Badiru & Marlin U. Thomas*

Industrial Control Systems: Mathematical and Statistical Models and Techniques, *Adedeji B. Badiru, Oye Ibidapo-Obe, & Babatunde J. Ayeni*

Industrial Project Management: Concepts, Tools, and Techniques, *Adedeji B. Badiru, Abidemi Badiru, & Adetokunboh Badiru*

Inventory Management: Non-Classical Views, *Mohamad Y. Jaber*

Kansei Engineering - 2 volume set

- Innovations of Kansei Engineering, *Mitsuo Nagamachi & Anitawati Mohd Lokman*
- Kansei/Affective Engineering, *Mitsuo Nagamachi*

Knowledge Discovery from Sensor Data, *Auroop R. Ganguly, João Gama, Olufemi A. Omitaomu, Mohamed Medhat Gaber, & Ranga Raju Vatsavai*

Learning Curves: Theory, Models, and Applications, *Mohamad Y. Jaber*

Modern Construction: Lean Project Delivery and Integrated Practices, *Lincoln Harding Forbes & Syed M. Ahmed*

Moving from Project Management to Project Leadership: A Practical Guide to Leading Groups, *R. Camper Bull*

Project Management: Systems, Principles, and Applications, *Adedeji B. Badiru*

Quality Management in Construction Projects, *Abdul Razzak Rumane*

Social Responsibility: Failure Mode Effects and Analysis, *Holly Alison Duckworth & Rosemond Ann Moore*

Statistical Techniques for Project Control, *Adedeji B. Badiru & Tina Agustiady*

STEP Project Management: Guide for Science, Technology, and Engineering Projects, *Adedeji B. Badiru*

Sustainability: Utilizing Lean Six Sigma Techniques, *Tina Agustiady & Adedeji Badiru*

Systems Thinking: Coping with 21st Century Problems, *John Turner Boardman & Brian J. Sauser*

Techonomics: The Theory of Industrial Evolution, *H. Lee Martin*

Triple C Model of Project Management: Communication, Cooperation, Coordination, *Adedeji B. Badiru*

FORTHCOMING TITLES

Essentials of Engineering Leadership and Innovation, *Pamela McCauley-Bush & Lesia L. Crumpton-Young*

Project Management: Systems, Principles, and Applications, *Adedeji B. Badiru*

Technology Transfer and Commercialization of Environmental Remediation Technology, *Mark N. Goltz*

SUSTAINABILITY

Utilizing Lean Six Sigma Techniques

Edited by
Tina Agustiady
Adedeji B. Badiru

CRC Press is an imprint of the
Taylor & Francis Group, an **informa** business

CRC Press
Taylor & Francis Group
6000 Broken Sound Parkway NW, Suite 300
Boca Raton, FL 33487-2742

Version Date: 20121105

International Standard Book Number: 978-1-4665-1424-9 (Hardback)

Visit the Taylor & Francis Web site at
http://www.taylorandfrancis.com

and the CRC Press Web site at
http://www.crcpress.com

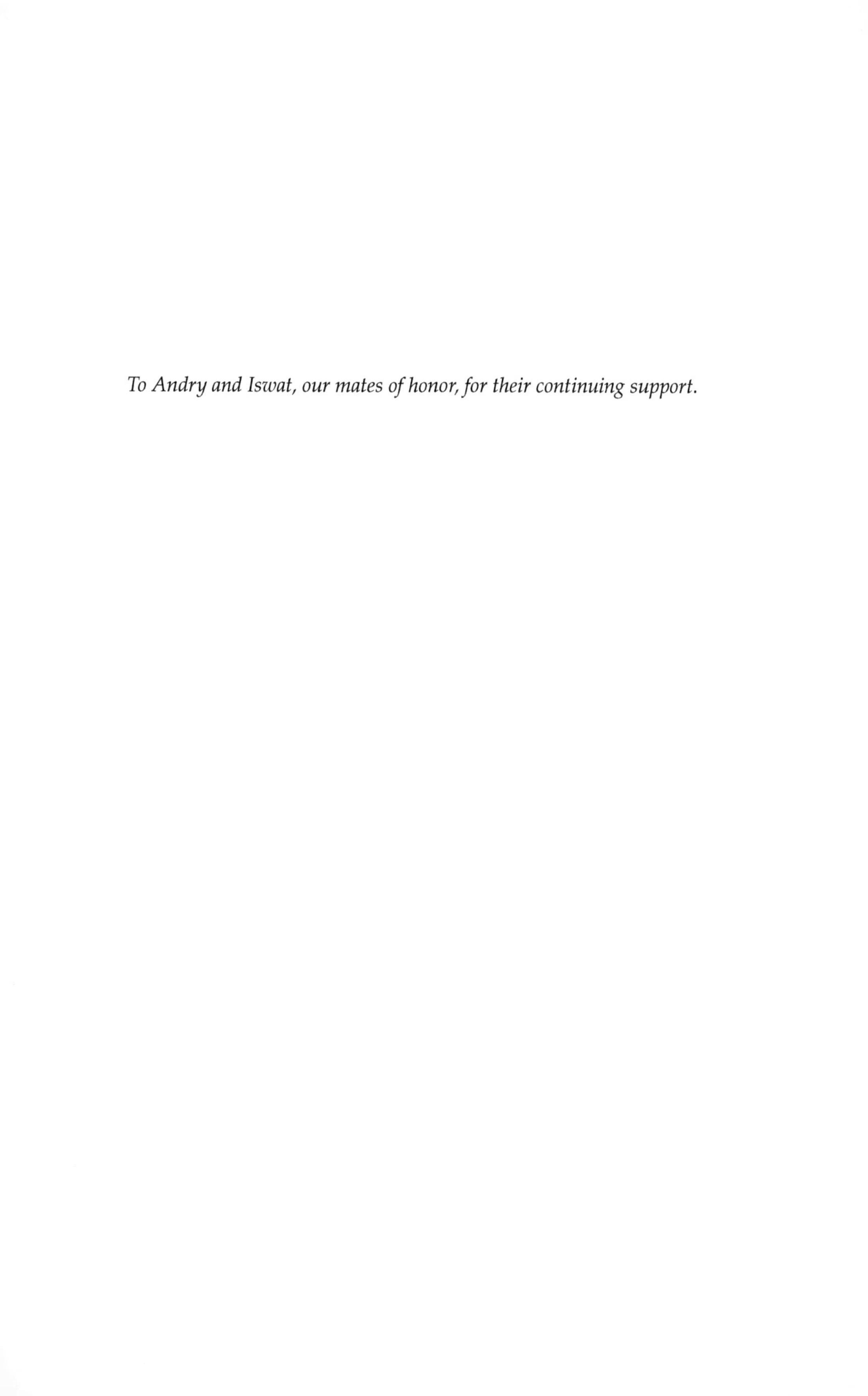

To Andry and Iswat, our mates of honor, for their continuing support.

Contents

Preface

This book presents the application of Six Sigma methodology as a viable path for achieving the goals of sustainability. Sustainability is a tool that helps companies gain true benefits from their improvements. Many companies make improvements but cannot keep the results successful. Tools such as Lean Six Sigma will help sustain results by using process-focused decisions. Examples of manufacturing improvements occur when companies implement 5S for organization before events but then end up getting disorganized just weeks after they have passed. These tools will help sustain the results while being competitive in this ever-changing economy. Manufacturing companies are working endlessly to make improvements, which are hard to implement and even harder to sustain. It is important to sustain best practices in order to not go back to the old ways of doing things and to be competitive in today's market. Implementation of sustainability is only successful with the proper tools and management.

The Lean and Six Sigma tools available can make sustaining results easier and process and business results more focused. The major features of the book include sustainability and metrics, Lean manufacturing, Six Sigma tools, sustainability project management, sustainability modeling, sustainable manufacturing and operations, decision making, and sustainability logistics. A key benefit of the book is that it will guide readers on how to sustain results from high-quality improvements to make organizations competitive and first-in-class in their respective lines of business. Benefits are always needed in companies for them to be successful and profitable. While continuous improvement techniques are sought after, the implementation of the techniques becomes difficult and challenging to sustain. The proper tools are the key to making the decisions and having successful results. Sustainability requires project management and accountability along with matrices to understand the order of events. To sustain results, simply relying on an individual or one particular database or tool is just not enough. True statistical techniques need to be implemented to help make each industry the best at what it does. Lean and Six Sigma are important tools that are here to stay. The tools are not only statistics but also show data-driven decisions and will implement results based on facts. Without utilizing these tools and leading this change, companies will become less and less marketable and profitable. Companies need sustainability utilizing Lean and Six Sigma techniques to be the best they can be.

Tina Agustiady
Adedeji B. Badiru

About the Authors

Tina Agustiady is a certified Six Sigma Master Black Belt and Continuous Improvement Leader at BASF. Agustiady serves as a strategic change agent, infusing the use of Lean Six Sigma throughout the organization as a key member of the site leadership team. She improves cost, quality, and delivery at BASF through her use of Lean and Six Sigma tools while demonstrating improvements through a simplification process. Agustiady has led many Kaizen, 5s, and root cause analysis events throughout her career in the healthcare, food, and chemical industries.

She has conducted training and improvement programs using Six Sigma for the baking industry at Dawn Foods. Prior to working at Dawn Foods, she worked at Nestlé Prepared Foods as a Six Sigma product and process design specialist responsible for driving optimum fit of product design and current manufacturing process capability and reducing total manufacturing cost and consumer complaints.

Agustiady received a BS in industrial and manufacturing systems engineering from Ohio University. She earned her Black Belt and Master Black Belt certifications at Clemson University. Agustiady is also the Institute of Industrial Engineers (IIE) Lean Division board director and chairperson for IIE annual conferences and Lean Six Sigma conferences. She is an editor for the *International Journal of Six Sigma and Competitive Advantage.* Agustiady is an instructor who facilitates and certifies students for Lean and Six Sigma for IIE and Six Sigma Digest. She is also the coauthor for *Statistical Techniques for Project Control* (published in January 2012) and *Communication for Continuous Improvement Projects* (to be published in May 2013).

Adedeji B. Badiru is the head of systems and engineering management at the Air Force Institute of Technology. He was previously the department head of industrial and information engineering at the University of Tennessee in Knoxville and formerly a professor of industrial engineering and the dean of University College at the University of Oklahoma. Dr. Badiru is a registered professional engineer (PE). He is a fellow of the Institute of Industrial Engineers and a Fellow of the Nigerian Academy of Engineering. He holds a BS in industrial engineering, an MS in mathematics, an MS in industrial engineering from Tennessee Technological University, and a PhD in industrial engineering from the University of Central Florida. His areas of interest include mathematical modeling, project modeling and analysis, economic analysis, systems engineering, and productivity analysis and improvement. Dr. Badiru is the author of several books and technical journal articles. He is the editor of the *Handbook of Industrial and Systems Engineering* and coeditor of the *Handbook of Military Industrial Engineering.* He is a member of several professional associations, including the Institute of Industrial Engineers (IIE), the Institute of Electrical and Electronics Engineers (IEEE), the Society of Manufacturing Engineers (SME), the Institute for Operations Research and Management Science (INFORMS), the American Society for Engineering Education (ASEE), the American Society for Engineering Management (ASEM), the New York Academy of Science (NYAS), and the Project Management Institute (PMI). He has served as a consultant to several organizations around the world, including Russia, Mexico, Taiwan, Nigeria, and Ghana. He has conducted customized training workshops for numerous organizations, including Sony, AT&T, Seagate Technology, the U.S. Air Force, Oklahoma Gas & Electric, Oklahoma Asphalt Pavement Association, Hitachi, Nigeria National Petroleum Corporation, and ExxonMobil. Dr. Badiru has won several awards for his teaching, research, publications, administration, and professional accomplishments.

1

Global Issues in Sustainability

Sustainability—The human preservation of the environment, whether economically or socially through responsibility, management of resources, and maintenance utilizing support.

Being sustainable in manufacturing and any financial institution only increases sales and profitability by reducing costs. Doing it is easy and involves awareness and education, policies and programs, and strategic planning.

Critical thinking by using Safety and Practicality techniques can transform our communities and global environments while moving into the next centuries. Sustainability is based on survival and well-being, which revolves around commonsense principles. Understanding our environment and establishing better practices utilizing continuous improvement principles will help us economically and help the generations to come. The topic sounds overwhelming, but utilizing Lean Six Sigma principles will make it a universal way of improving our lives.

When determining what to do to be sustainable, four main concepts should be sought after:

- Economical
- Personal
- Societal
- Environmental

Economical terms are related to profitability in the economy. Personal means of sustainability are the effects that are subjected to an individual. Societal concepts deal with informal social gatherings of groups organized by something in common. Finally, the environment is the setting, surroundings, or conditions in which living objects operate. These four concepts unite for sustainability because of the parallel relationship they have to one another.

The basic questions should be asked first:

- What needs to be done?
- What can be done?
- What will be done?
- Who will do it?
- When will it be done?

- Where will it be done?
- How will it be done?

The aspects that are critical to sustainability are the most basic ones. These include the environment, natural resources, energy, air, water, and the proper use of them, and finally the understanding of physics, chemistry, biology, and geology. As humans, we must be economical so that none of these resources become scarce, and human growth maintains predominance. Many may ask, "Why should I protect my environment, and why should I be economical?" The answer is simple: It is for well-being. This well-being can be of an individual, resources, work, proper living manners, and pure satisfaction. Without being sustainable, our environment will suffer from aspects such as pollution, not having enough renewable resources especially essential ones, and livelihoods.

The same aspects that are thought about in everyday life for our own well-beings should be thought about for work purposes as well. Emissions are powerful, and companies must reduce them to be successful. Some companies are mandated to only pollute a certain percentage into the environment. Recall that being sustainable in manufacturing only increases sales and profitability by reducing costs. The quality of products is always increased when sustainability is built in. Sustainability also means predictability or doing the same right thing every time. When costs are cut, quality should not be altered or the increase in sales will decrease drastically. The processes behind the quality are the most critical ones to be sustainable. It should be sought after to have value-added tasks with quality products. When customers see that sustainability is not being met, they will go elsewhere. This also happens when the customers have expectations and are surprised with the product each different time they purchase the product. One time the product met their expectations, so the customers came back. Every time they receive a different product than they expect, the customers get disappointed and turn to a different supplier for their product. Not only do the customers go elsewhere, but they complain to their friends and family about their dissatisfaction. Their friends and family now complain to other friends and family about the story they heard, and the company loses sales—hence profitability—by sacrificing their quality.

The main global issue in companies is the wastes they incur. The eight main wastes are the following:

Transportation
Inventory
Motion
Waiting
Overprocessing
Overproduction

Defects/rework
Underutilization of employees

Each of the categories is discussed below.

Transportation

Materials or parts transferred within the facility should be minimized in the most strategic manner possible. If a part is coming from shipping and going to manufacturing, the shipping dock must be as close to the manufacturing dock as possible. When each part is utilized, the parts being put in the product should all be in similar vicinities so there is no waste in transporting the parts back and forth. Also, when the part is completed with its assembly or manufacturing, the packaging area and storage area should be close by or an ease of transportation must occur to limit this type of waste.

Inventory

Inventory should always be looked at as dollar signs on the shelves of where the inventory sits. The raw materials, in-process items, and finished goods are not value-added items when just sitting in a location. Just-in-time (JIT) methodologies should be used to reduce having excess inventory.

JIT is a value-streaming methodology created by Taiichi Ohno at Toyota in the 1950s. The production and delivery of proper items and proper times is the philosophy for JIT to be successful. Upstream activities must occur before downstream activities, aiming for single-piece flow.

Flow, pull, standard work, and Takt are the key elements. Changeovers need to be minimized in order for upstream manufacturing processes to have minimal parts until the downstream manufacturing process is ready. This prevents accumulation of excess stock. Scheduling must be effective and consistent for this process to be maintained. Overextending production will also cause inventory issues. Demand of customers needs to be measured before producing.

Flow is looked at to enhance efficiency. Value-added tasks should be looked for even when difficult to analyze. Many employees find tasks such as cleaning to be value added when in fact they are not. The only tasks that add value are the tasks that the customer would be willing to pay for. The focus should be on the end product and what the customer wants. If the customer is buying a chocolate cake with chocolate frosting, they only want to pay for this cake. They do not want to pay for the changeover between strawberry and chocolate that has to occur before they get their chocolate cake.

Standardized work, also known as standard operating procedure (SOP), consists of a definitive set of work procedures with all tasks organized optimally to meet customer needs. Standardization of the process and the activities allow for consistent times and completion of entire processes by all employees at all times and all circumstances.

Correct locations of materials, parts, equipment, and so on, also help flow of materials and processes. 5S is a key tool used when determining flow. 5S will be discussed further in the book.

Motion

Any wasted time performing any activity is considered a waste of motion. The topic is similar to what was discussed earlier. The customer only wants to pay for value-added tasks. The customer does not want to pay excess labor fees, which are inherent when an employee is unorganized and looking for materials. Any type of wasted motion such as double-handling ingredients, walking further than needed, and stacking multiple times are considered wastes of motion.

Waiting

When any person is waiting for tools, materials, equipment, or other personnel, they are being non-value-added. JIT should again be used, yet this time with people as the driving factor. The next step in the process should come just in time for the next person to be able to act on that process without wait time occurring. If wait time is built into the process, there is excess time built into the process.

Overprocessing

A question many people ask when doing tasks is, "Why do you do it that way?" The common answer is, "That's how we have always done it," or "That's what the procedure says." What should be thought about is if the process has too much work built into it. If a process requires 5 minutes of stirring, it should be asked if the materials would be the exact same after 3 minutes of stirring. Minimizing excess processing saves time and eliminates waste while increasing efficiency and productivity. This type of thinking requires "thinking outside of the box."

The strategy for utility cost reduction is the following:

- Utility Cost = Utility Usage × Utility Price
- Electricity $/Year = kWh/Year × $/kWh
- Natural Gas $/Year = Cu. Ft./Year × $/Cu. Ft.
- Water $/Year = Gallons/Year × $/Gallon

Simply put, utility cost reductions are the product of utility usage reductions and utility price reductions. Focusing entirely on usage reductions or entirely on price reductions is not a well-balanced strategy.

To correctly estimate utility usage and cost savings, two estimates must be prepared. Annual utility usage savings and cost savings are the difference between the "before" usage/cost estimate and the "after" usage/cost estimate:

- Annual "Before-Retrofit" Utility Usage/Cost – Annual "After-Retrofit" Utility Usage/Cost = Annual Utility Usage/Cost Savings

Because this is a multifactor analysis (e.g., before and after levels of equipment loading/efficiency/demand, before and after annual hours of operation, before and after utility prices), there are lots of opportunities for errors. Thus, the uncertainty of the utility usage and cost reduction estimates must be taken into account. The conservative approach is to discount the estimates by the associated level of uncertainty.

The following are utility cost reduction measures and focuses for sustainability:

1. Efficient operation and effective maintenance of utility-consuming equipment
2. Competitive procurement of utilities
3. Cost-effective expense improvements
4. Cost-effective capital projects and retrofits
5. Efficient design of new buildings and plant expansions

The following are examples through simple means of sustainability:

- Computers, radios, and stereo systems should be turned *off* when not in use and unplugged when possible.
- Utilize programmable thermostats to save up to 20% annually.

- Turn off electricity in rooms not being utilized—Think about freezers in factories that are not being used.
- Focus on:
  - Electricity—#1 source of resource consumption
  - Natural gas—#2 source of resource consumption
  - Water/sewer—#3 source of resource consumption

Utility costs are one of the largest categories in the annual expense budget for facilities and maintenance organizations. Unlike property taxes and depreciation costs, utility costs can be readily reduced.

Preventative maintenance is 67% to 75% less expensive than repair-upon-failure, so it's a measure that reduces net expenses concurrent with implementation. Preventative maintenance need not be applied wholesale—it can be implemented in stages and/or for selected categories of assets.

The eight best practices for improved energy efficiency are the following:

1. Increase the efficiency of all motors and motor-driven systems
2. Improve building lighting
3. Upgrade heating, ventilating and cooling systems
4. Capture the benefits of utility competition
5. Empower your employees to do more
6. Use water-reduction equipment and practices
7. Explore energy savings through increased use of the Internet
8. Implement comprehensive facility energy and environmental management

The next topic discusses the areas of sustainability with particular aspects.

Areas of Sustainability

When discussing the different areas of sustainability, particular aspects come to mind. Sustaining the environment is first and foremost, but we need to understand what is included in the environmental aspects. The most common ways to sustain the environment is through energy, water, and agricultural conservation. To do this, we must be well educated in sustainability for it to be implemented and have lasting results.

First off, we need to understand energy consumption. Energy is defined as the capacity of a physical system to perform work through heat, kinetics, mechanical systems, light, or electrical means.

The beginning process of saving energy comes from simple means such as turning off lights and any sources that provide power. Computers, radios, stereo systems, and so on, should be not only turned off when not in use but also unplugged when possible. Surge protectors are good ways of being able to ease into this process. Multiple items can be plugged into the surge protector and the switch can simply be turned off when not in use.

Many thermostats are programmable now so that the air-conditioning or heat can be turned to a lower/higher temperature when you are off at work or elsewhere. The trick to a programmable thermostat is that it gradually gets back to the temperature of normal operating conditions when you reenter the home. These savings can be up to 20% annually if utilized properly. Another great way to save energy in your home is to close any vents to rooms that are underutilized. The extra heat or air will come into the other rooms, saving costs again.

Weatherproofing windows with weather strips or plastic insulation will also help prevent drafts or leaks. Curtains can also help with this process and insulated curtains are currently available to keep in heat or air. Simply insulating homes can protect excess heat or air costs and has a quick payback of normally less than one year.

Microwaves or small toaster ovens use much less energy than stoves or ovens. These should be used when cooking or defrosting small amounts of food.

Keeping the cold energy in the refrigerator is important because of the amount of energy it uses. Therefore closing the fridge as soon as possible is very important. Some fridges come with energy savings signals such as beeps to warn you if the fridge has been open for longer than the allotted time. Remember that old fridges also account for a large amount of energy.

Batteries also take up energy that can be saved by turning off any toys, games, and so on. Rechargeable batteries can save energy and also reduce the toxins of the dangerous heavy metals that are emitted from batteries when thrown away. Batteries should only be disposed in toxic waste disposals.

Water is the next critical area where sustainability is considered. The basics of turning off water whenever possible are the key to the savings. Most cities use the most amount of energy supplying water and cleaning up the water after it has been used. Turning off water when brushing teeth or when washing dishes are simple ideas. Taking shorter showers is another way to save on the use of hot water. Hot water heaters account for almost ¼ of the home's energy usage.

Surprisingly, dishwashers save up to 40% more water than hand-washing dishes as well! If hand washing is a necessity, the best way to save is to fill up one side of the sink with clean soapy water and rinsing on the other side.

Using your own energy versus power sources will help in energy savings as well. Examples such as using rakes versus leaf blowers or push mowers versus gas mowers not only save energy but also help your own well-being!

Anything that can be thrown away should be thought about before purchasing. Think about disposable products such as plastic cups, plastic bags, napkins, and paper plates. Whenever a reusable source can be used, it should. Cloth napkins can cut down on the overuse of paper napkins. Also, the plastic bags given at grocery stores normally end up in the trash. Consider taking your own bags with you and reusing them. Newspapers should always be recycled; 36 million trees a year could be saved if everyone recycled their newspaper every day.

If these simple things can be done quickly and efficiently in one's home, think about the benefits in a corporation for these savings. The benefits can have major savings and the paybacks for buying energy-efficient bulbs, surge protectors, automated light sensors, and so on, will be immediate.

The beginning programs for these sustainability acts are simple. They consist of the following:

- Awareness
- Energy-efficient coding
- Communications programs
- Green schools and offices programs
- Energy alliances programs
- Industrial programs
- Policy and research programs
- Water savings acts
- Commercial buildings consortiums

Awareness begins with education. Public awareness, training, and classroom activities will help the society move toward sustainable actions for the environment. Attainability becomes an argument when discussing sustainability. The lack of agreement behind these acts should be followed up with facts behind sustainability and will inform others of truths behind the matter. Instead of focusing on what is wrong with our society and how difficult it is to maintain these sustainable measures, we should begin with the quick fixes and how simple ideas can help the environment. If each person starts doing just a small part, the effects can be immense. The small measures will then lead to much larger projects, but the effort has to begin somewhere. The big differences between the reality of being sustainable as a culture and the education behind being sustainable should be known. Education will lead the environment from being protected in small pieces while slowly investing in the ideas.

Education will affect three main areas of planning:

- Implementation
- Decision making
- Quality of life

Without educating the workforce to keep them informed, the implementation of sustainability cannot be founded. Highly illiterate forces have a difficult time with developmental plans and actions. Instead these types of areas and people will have to buy forms of energy and goods, spending a great deal of money.

Decision making is important because without factual and motivational information, decisions cannot be made. Development options with educational topics will help people be skilled and technical and will also lead the act of persuasion. The persuasion is simple due to having educational information that leads to a better society because the facts protect the environment and social structure.

Finally, the quality of life is important to each individual person. The learning of these practices leads to higher economic statuses, improvement of life conditions, and the future of coming generations. The quality of life helps both individuals and national geography.

Sustainable development plays a key role during this educational phase. The concept behind sustainable development is still evolving. The definition of sustainable development is the concept of needs mostly from the idea of limitations imposed by society and technology to prevent the environment from meeting present and future needs. The Brundtland Commission is to be credited for the original description of sustainable development. They stated, "Sustainable development is development that meets the needs of the present without compromising the ability of future generations to meet their own needs."

This concept is thought to have three main components:

- Environment
- Society
- Economy

These acts need to all be combined and not seen separately. The thought is to have a balance of all of these to have a healthy environment with proper resources for its citizens in the cleanliest manner possible.

Remember that the environment is the setting, surroundings, or conditions in which living objects operate. Society is the collective number of people living together in a particular region with the same customs, laws, or organizations. The economy is the prosperity, possessions, and resources of a country or region in terms of production and consumption of goods and services. They can all be tied together because they depend on each other.

Many principles are tied in with sustainable development. The definitions have been established from *The Rio Declaration on Environment and Development* and the 18 points are listed below:

- People are entitled to a healthy and productive life in harmony with nature.
- Development today must not undermine the development and environment needs of present and future generations.

- Nations have the sovereign right to exploit their own resources, but without causing environmental damage beyond their borders.
- Nations shall develop international laws to provide compensation for damage that activities under their control cause to areas beyond their borders.
- Nations shall use the precautionary approach to protect the environment. Where there are threats of serious or irreversible damage, scientific uncertainty shall not be used to postpone cost-effective measures to prevent environmental degradation.
- In order to achieve sustainable development, environmental protection shall constitute an integral part of the development process, and cannot be considered in isolation from it. Eradicating poverty and reducing disparities in living standards in different parts of the world are essential to achieve sustainable development and meet the needs of the majority of people.
- Nations shall cooperate to conserve, protect, and restore the health and integrity of the Earth's ecosystem. The developed countries acknowledge the responsibility that they bear in the international pursuit of sustainable development in view of the pressures their societies place on the global environment and of the technologies and financial resources they command.
- Nations should reduce and eliminate unsustainable patterns of production and consumption, and promote appropriate demographic policies.
- Environmental issues are best handled with the participation of all concerned citizens. Nations shall facilitate and encourage public awareness and participation by making environmental information widely available.
- Nations shall enact effective environmental laws, and develop national law regarding liability for the victims of pollution and other environmental damage. Where they have authority, nations shall assess the environmental impact of proposed activities that are likely to have a significant adverse impact.
- Nations should cooperate to promote an open international economic system that will lead to economic growth and sustainable development in all countries. Environmental policies should not be used as an unjustifiable means of restricting international trade.
- The polluter should, in principle, bear the cost of pollution.
- Nations shall warn one another of natural disasters or activities that may have harmful transboundary impacts.
- Sustainable development requires better scientific understanding of the problems. Nations should share knowledge and innovative technologies to achieve the goal of sustainability.

- The full participation of women is essential to achieve sustainable development. The creativity, ideals, and courage of youth and the knowledge of indigenous people are needed too. Nations should recognize and support the identity, culture, and interests of indigenous people.
- Warfare is inherently destructive of sustainable development, and nations shall respect international laws protecting the environment in times of armed conflict, and shall cooperate in their further establishment.
- Peace, development, and environmental protection are interdependent and indivisible.

These "Rio Principles" are clear parameters for the vision of the future of the world in order to be developmentally sustainable with specific abstracts of factual concepts.

Energy-efficient coding comprises the codes that set minimum requirements for energy-efficient design and construction for new and renovated buildings that impact energy usage. The objective is to have consistent and long-lasting results for energy savings. The benefits will help the environment both tangibly and intangibly. The reduction in energy helps the pollution concept by reducing environmental pollutions. There are direct savings and financial benefits due to a decrease in energy consumption and an increase in energy-efficient technologies, which will also lead to economic opportunities for the businesses. Once a building is already constructed, it is difficult to achieve the energy efficiencies without retrofitting the building. Since buildings stay up for decades, the energy coding should be done at the beginning stages for maximum benefits.

Contemporary energy codes can save up to 330 trillion BTUs by 2030, which is almost 2% of total current residential energy consumption. The states can take the lead in the energy-saving initiatives by mandating energy coding in new or retrofitted buildings (data from the U.S. Department of Energy).

Communications programs for sustainability include basic marketing techniques. Commercials, T-shirts promoting energy-saving activities, and mobile media are the most common ways to spread the word. Programs include seminars to raise awareness and scholarship grants to reward students with bright ideas. Finally donations help the sustainable efforts for the communications plans for the future.

Green schools and offices programs comprise of many communities that take initiatives to have only green products. The communication of where and what to get to be green is communicated via directories and online sources. Green schools and offices programs give hazardous material warnings of items such as PVC plastic and give the benefits of recycling programs. These programs are meant to show the community their commitment in the matter by safeguarding health, saving money, raising test scores through better air quality and environments, and empowering kids.

Energy alliance programs give tips and resources for making energy efficiency easy and affordable. There are many community-based nonprofit organizations that provide property owners with information on what to do, who to call, how to pay for resources, and what the proper tactics are for energy savings. Programs such as these are not associated with any political parties and are solely performed for better environments. The programs have also begun to involve renters, contractors, and business partners.

Industrial programs perform similar techniques for energy savings by having steering committees to gain executive buy-in from the beginning and create sustainable goals and establishments with measurable results. Industrial programs involve best practices by exploring company motivations for the program that have had successive results. The marketing research is a key aspect to the successes for industrial programs. The goal of industrial programs is to have clean air and transportation, upcoming technology, corporate sustainability, and a means for building efficiency sectors.

Policy and research programs involve sustainable science programs by advancing science through awareness of human-environment systems. The knowledge behind the policies increases the ease of implementation by having data-driven results for the changes to be implemented. The promotion for sustainability includes concepts to increase knowledge, train students and faculty, and provide continuous education through teaching and outreach.

There are several water savings acts that involve goals for reduction in urban water usage by 20% in 8 to 10 years. Incremental changes are to be taken with a goal of having at least 10% of savings in the next 5 years. Water management plans encompass giving deliverables on the amount of water usage acceptable through technical methodologies with strict criteria. The water savings act has a goal to not approve any retail water suppliers for any grants or loans if their water conservation requirements are not met. There are plans for an adoptive pricing structure for water by basing it on parts or quantities delivered. Efficient management strategies are to be encompassed in this water savings act.

Commercial buildings consortiums are developed by the U.S. Department of Energy to gain building asset rating systems that can be built for new and existing commercial buildings to provide information to the stakeholders. The concept is due to the loss of heating and cooling by up to 30% from commercial buildings through the poor use or insulation of doors, windows, curtains, and skylights, which is also known as fenestration. The goal behind commercial building consortiums is to have all building sectors in the United States utilize zero energy through the implementation of aggressive energy savings mechanisms to reduce demand of energy by 70 to 80% while still maintaining energy requirements through renewable resources.

These different areas of sustainability help benchmark the future use of energy by primarily using the domination of communication, knowledge, and education as a means of securing the well-being of future generations.

What Happens Globally if We Are Not Sustainable?

According to the U.S. Census, there are approximately 312 million people in the United States and approximately 6.9 billion people in the world. With this many people, there is a fear of running out of natural resources, especially energy. Renewable sources of energy are needed to be sustainable. Ultimately any living organism requires energy, and if all living organisms utilize energy without reproducing it in some way, ultimately the energy *will* run out. Revitalizing all forms of energy must be accomplished to be sustainable. This means reorganizing, restoring, and differentiating. If we continue to use resources without reinvesting, all energy will be lost and there will be no hope for future generations. Social beings working together can help current and future generations by sustaining positive relationships through education and understanding of laws and practices. Locally purchasing ingredients and goods also helps the environment so there are no monopolies. Monopolies are defined as one company or industry controlling the services or goods provided. With the concept of monopoly, no other beings are given a chance to be held socially responsible. The consequences include not having hope for future generations to be successful because no bartering can be done from one business or organization to another.

Overconsumption can come from pressures of being known as successful or prestigious. However, having the best cars, homes, and things will lead to holding others at a pace they cannot keep up with. According to "The Story of Stuff" (2007), we have consumed 30% of Earth's natural resources in the past 30 years. Only a few of these resources can be replenished. Only 20% of old-growth forest is remaining and 75% of fisheries are producing at or above capacity. The United States is the most abusive global consumer with only 5% of the world population and 30% of the worldwide consumption. Ninety-nine percent of raw materials are discarded without being recycled. This will be catastrophic if these standards are maintained or worsened over the years.

The food chain circumstance that animals use can be converted to humans with this mannerism, meaning only the best or most successful will come out on top. Instead all philosophies should be investigated before making decisions. Values can then be set with the proper principles in mind, holding relationships as important aspects. This is another form of unselfishness that should be practiced to stay globally friendly and sustainable. These concepts will not only help current practices and environments but also provide even better hopes for future generations. If we keep consuming resources and just moving to different areas to find more resources, eventually the resources in all areas will run out, leaving the world completely out of resources.

References

Commercial Buildings Initiative. (2008). Zero Energy Commercial Buildings Consortium. *Energy Efficiency Toolkit for Manufacturers: Eight Proven Ways to Reduce Your Costs.* Accessed from http://www.energy.ca.gov/process/pubs/toolkit.pdf.

Fiorino, D. P. (2004). Case study: *Utility Cost Reduction at a Large Manufacturing Facility.* Accessed from http://texasiof.ces.utexas.edu/PDF/Presentations/Nov4/CaseStudyFiorino.pdf.

Green Schools Initiative, The David Brower Center.

LEAP. (2010). Local Energy Alliance Program. Charlottesville, VA.

"The Story of Stuff." (2007). http://www.storyofstuff.com/.

Water Use Efficiency Branch. (2010). Sacramento, CA.

2

Systems View of Sustainability

Sustainability implies repeatability, the ability to replicate success again and again to create an enduring pattern of achieving the end goal. Repeatability is not limited to the attributes of the end product. It must also be pursued and practiced in all facets of operations from a systems perspective. It is only through a systems view that we can address the multitude of issues and factors associated with sustainability.

What Is a System?

A system is a collection of interrelated elements working together synergistically to achieve a set of objectives. Any project is essentially a collection of interrelated activities, people, tools, resources, processes, and other assets brought together in the pursuit of a common goal. The goal may be in terms of generating a physical product, providing a service, or achieving a specific result. This makes it possible to view any project as a system that is amenable to all the classical and modern concepts of systems management.

A systems view of the world makes sustainability work better and projects more likely to succeed. A systems view provides a disciplined process for the design, development, and execution of complex sustainability initiatives in business, industry, education, and government.

A major advantage of a systems approach to sustainability is the win-win benefit for everyone. A systems view also allows full involvement of all stakeholders of sustainability.

A Systems Engineering Framework

Systems engineering is the application of engineering to solutions of a multifaceted problem through a systematic collection and integration of parts of the problem with respect to the life cycle of the problem. It is the branch of engineering concerned with the development, implementation, and use of large or complex systems. It focuses on specific goals of a system considering the specifications, prevailing constraints, expected services, possible behaviors, and structure of the system. It also involves a consideration of the activities required to ensure that the system's performance matches the stated goals. Systems engineering addresses the integration of tools, people, and processes required to achieve a cost-effective and timely operation of the system. We are all stakeholders of sustainability. As such, we are subcomponents of a large societal system. Systems engineering is concerned with the big picture of the challenges that we face. This involves how a system functions or behaves overall, how it interfaces with its users, how it responds to its environment and interacts with other systems, and how it regulates itself. Sustainability must be approached both from a technical point of view as well as a management viewpoint. The management discipline organizes and allocates efforts and resources across the broad spectrum of the system, including initiating communication, facilitating collaboration, defining system requirements, planning work flows, and deploying technology for targeted needs. A systems engineering framework for sustainability has the following elements:

- Focus on the end goal.
- Involve all stakeholders.
- Define the sustainability issue of interest.
- Break down the problem into manageable work packages.
- Connect the interface points between project requirements and project design.
- Define the work environment to be conducive to the needs of participants.
- Evaluate the systems structure.
- Justify every major stage of the sustainability project.
- Integrate sustainability into the core functions existing in the organization.

Definitions of Sustainability

Different experts, researchers, and practitioners have differing definitions of sustainability. The lack of a consistent view is probably the reason that many sustainability programs have not taken root as expected. The following systems-based definitions are essential for the purpose of the theme of this book:

Agriculture—Farming and manipulation of soils to harvest and grow crops while raising livestock

Biodiversity—The natural variation of living in particular ecosystems

Biophysical—The science dealing with the application of physics of biological processes

Biosphere—The actual regions where living organisms occupy or reside

Climate—Temperature, precipitation, and wind characteristics and conditions

Community service—Voluntary work intended to assist others in a particular area

Cooperation—The process of two or more beings working together toward the same goal

Coordination—Organization of different elements so that there is cooperation for effective processes

Economy—The prosperity, possessions, and resources of a country or region in terms of production and consumption of goods and services

Ecosystem—Interacting organisms and their physical means of living in a biological community

Education—Learning or training achieved through knowledge and studying

Energy—The capacity of a physical system to perform work through heat, kinetics, mechanical systems, light, or electrical means

Energy-efficient coding—Codes that set minimum requirements for energy-efficient design and construction for new and renovated buildings that impact energy usage

Enterprise—A business or company with resources

Environment—The setting, surroundings, or conditions in which living objects operate

Ethics—A system of moral values that makes one perform the right conduct

Fenestration—The arrangement of windows and doors in a building to help operate with lower heating and cooling losses

Forecasting—The prediction or estimation of future events performed normally due to trending

Global—Relation to the world, earth, or planet

Global warming—The gradual increase in temperature of the earth and oceans predicted to be from pollution and the inconsideration of the environment

Greenhouse—Solar radiation entrapment caused by atmospheric gases caused by pollution, which allows sunlight to pass through and be re-emitted as heat radiation back from the earth's surface

Human impact—The impacts from human beings that affect biophysical environments, biodiversity, and any other environments

Humanity—Human nature and civilization

Natural resources—Materials or matter in nature such as minerals, fresh water, forests, or abundant land that can be used for economic benefits

Physics—The science of matter and energy and the interaction of the two

Planetary—Relating to the earth as a planet

Policies—Plans or courses of actions

Project management—Planning, managing, monitoring, and controlling projects utilizing feedback and knowledge of tools and techniques

Public health—The science and art of preventing diseases while prolonging life

Quality management—The key for a system to ensure the end goal is met through a desired level of excellence to be competitive in the business

Recycling–The reprocessing of materials already used into new materials or products to prevent waste and protect the environment by reducing energy, air pollution, water pollution, and emissions

Reducing waste—Reducing the amount of unwanted materials technologically and socially to economically benefit the environment

Reuse—To use again in a different circumstance after processing

Society—The collective amounts of people living together in a particular region with the same customs, laws, or organizations

Sustainability—The human preservation of the environment, whether economically or socially through responsibility, management of resources, and maintenance utilizing support

Sustainable development—The concept of needs mostly from the idea of limitations imposed by society and technology to prevent the environment from meeting present and future needs

Sustainable energy—The condition of energy that meets the needs of the present without compromising the ability of future generations to meet their needs

Systems—Detailed methods or procedures established to transmit out a specific activity or to perform a responsibility; a network of things interacting together.

Theory—A set of assumptions based on accepted facts that provide rational explanations of cause-and-effect relationships among groups

Total quality management—Continuous improvement in products and processes by increasing the quality and reducing the defects through management methodologies

The Many Languages of Sustainability

Sustainability is not just for the environment. Although environmental concern is what immediately comes to mind whenever the word *sustainability* is mentioned, there are many languages (i.e., modes) of sustainability, depending on whatever perspective is under consideration. The context determines the interpretation. Each point of reference determines how we, as individuals or groups, respond to the need for sustainability. Pursuits of green building, green engineering, clean water, climate research, energy conservation, eco-manufacturing, clean product design, lean production, and so on remind us of the foundational importance of sustainability in all we do.

What Is Sustainability?

Commitment to sustainability is in vogue these days, be it in the corporate world or personal pursuits. But, what exactly is sustainability? Definitions of the word contain verbs, nouns, and adjectives such as *green, clean, maintain, retain, stability, ecological balance, natural resources,* and *environment*. The definition of *sustainability* implies the ability to sustain (and maintain) a process or object at a desirable level of utility. The concept of sustainability applies to all aspects of functional and operational requirements, embracing both technical and managerial needs. Sustainability requires methodological, scientific, and analytical rigor to make it effective for managing human activities and resources.

In the above context, sustainability is nothing more than prudent resource utilization. The profession of industrial engineering is uniquely positioned to facilitate sustainability, especially as it relates to the environment, technical resources, management processes, human interfaces, product development, and facility utilization. Industrial engineers have creative and simple solutions to complex problems. Sustainability is a complex undertaking that warrants the attention and involvement of industrial engineers. A good example of the practice of sustainability is how a marathon runner strategically expends stored energy to cover a long-distance race. Burning up energy too soon means that the marathon race will not be completed. Erratic expenditure of energy would prevent the body from reaching its peak performance during the race. Steady-state execution is a foundation for achieving sustainability in all undertakings where the decline of an asset is a concern. An example is an analysis of how much water or energy it takes to raise cattle for human consumption. Do we ever wonder about this as we delve into our favorite steaks? Probably not. Yet, this is, indeed, an issue of sustainability.

Resource Consciousness

The often-heard debate about what constitutes sustainability can be alleviated if we adopt the context of "resource consciousness," which, in simple terms, conveys the pursuit of conservation in managing our resources. All the resources that support our objectives and goals are amenable to sustainability efforts. For example, the expansion of a manufacturing plant should consider sustainability, not only in terms of increased energy consumption but also in terms of market sustainability, intellectual property sustainability, manpower sustainability, product sustainability, and so on. The limited resource may be spread too thin to cover the increased requirements for a larger production facility. Even a local community center should consider sustainability when contemplating expansion projects just as the local government should consider tax base sustainability when embarking on new programs. The mortgage practices that led to the housing industry bust in the United States were due to financial expectations that were not sustainable. If we put this in the context of energy consumption, it is seen that buying a bigger house implies a higher level of energy consumption, which ultimately defeats the goal of environmental sustainability. Similarly, a sports league that chooses to expand haphazardly will eventually face a non-sustainability dilemma. Every decision ties back to the conservation of some resource (whether a natural resource or a manufactured resource), which links directly to the conventional understanding of sustainability.

Foci of Sustainability

There are several moving parts in sustainability. Only a systems view can ensure that all components are factored into the overall pursuit of sustainability. A systems view of the world allows an integrated design, analysis, and execution of sustainability projects. It would not work to have one segment of the world embarking on sustainability efforts while another segment embraces practices that impede overall achievement of sustainability. In the context of production for the global market, whether a process is repeatable or not, in a statistical sense, is an issue of sustainability. A systems-based framework allows us to plan for prudent utilization of scarce resources across all operations. Some specific areas of sustainability include the following:

- Environmental sustainability
- Operational sustainability
- Energy sustainability
- Health and welfare sustainability
- Safety and security sustainability
- Market sustainability

- Financial sustainability
- Economic sustainability
- Health sustainability
- Family sustainability
- Social sustainability

The long list of possible areas means that sustainability goes beyond environmental concerns. Every human endeavor should be planned and managed with a view toward sustainability.

Value Sustainability

Sustainability imparts value on any organizational process and product. Even though the initial investment and commitment to sustainability might appear discouraging, it is a fact that sustainability can reduce long-term cost, increase productivity, and promote achievement of global standards. Sample questions for value sustainability are provided below:

- What is the organizational mission in relation to the desired value stream?
- Are personnel aware of where value resides in the organization?
- Will value assignment be on team, individual, or organizational basis?
- Is the work process stable enough to support the acquisition of value?
- Can value be sustained?

Using the Hierarchy of Needs for Sustainability

The psychology theory of hierarchy of needs, postulated by Abraham Maslow in his 1943 paper, "A Theory of Human Motivation," still governs how we respond along the dimensions of sustainability, particularly where group dynamics and organizational needs are involved. An environmentally induced disparity in the hierarchy of needs implies that we may not be able to fulfill personal and organizational responsibilities along the spectrum of sustainability. In a diverse workforce, the specific levels and structure of needs may be altered from the typical steps suggested by Maslow's hierarchy of needs. This calls for evaluating the needs from a multidimensional perspective. For example, a 3-D view of the hierarchy of needs can be used to coordinate personal needs with organizational needs with the objective of facilitating sustainability. People's hierarchy of basic needs will often dictate how they respond to calls for sustainability initiatives. Maslow's hierarchy of needs consists of five stages:

1. *Physiological needs*: These are the needs for the basic necessities of life, such as food, water, housing, and clothing (i.e., survival needs). This is the level where access to money is most critical. *Sustainability applies here.*
2. *Safety needs*: These are the needs for security, stability, and freedom from physical harm (i.e., desire for a safe environment). *Sustainability applies here.*
3. *Social needs*: These are the needs for social approval, friends, love, affection, and association (i.e., desire to belong). For example, social belonging may bring about better economic outlook that may enable each individual to be in a better position to meet his or her social needs. *Sustainability applies here.*
4. *Esteem needs*: These are the needs for accomplishment, respect, recognition, attention, and appreciation (i.e., desire to be known). *Sustainability applies here.*
5. *Self-actualization needs*: These are the needs for self-fulfillment and self-improvement (i.e., desire to arrive). This represents the stage of opportunity to grow professionally and be in a position to selflessly help others. *Sustainability applies here.*

Ultimately, the need for and commitment to sustainability boil down to each person's perception based on his or her location on the hierarchy of needs and level of awareness of sustainability. How do we explain to a hungry poor family in an economically depressed part of the world the need to conserve forestry? Or, how do we dissuade an old-fashioned professor from the practice of making volumes of hard copy handouts instead of using electronic distribution? Cutting down on printed materials is an issue of advancing sustainability. In each wasteful eye, "the *need* erroneously justifies the *means*" (author's own variation of the common phrase). This runs counter to the principle of sustainability. In Figure 2.1, this article expands the hierarchy of needs to generate a 3-D rendition that incorporates organizational hierarchy of needs. The location of each organization along its hierarchy of needs will determine how the organization perceives and embraces sustainability programs. Likewise, the hierarchy position of each individual will determine how he or she practices commitment to sustainability.

In an economically underserved culture, most workers will be at the basic level of physiological needs, and there may be constraints on moving from one level to the next higher level. This fact has an implication on how human interfaces impinge upon sustainability practices. In terms of organizational hierarchy of needs, the levels in Figure 2.1 are characterized as follows:

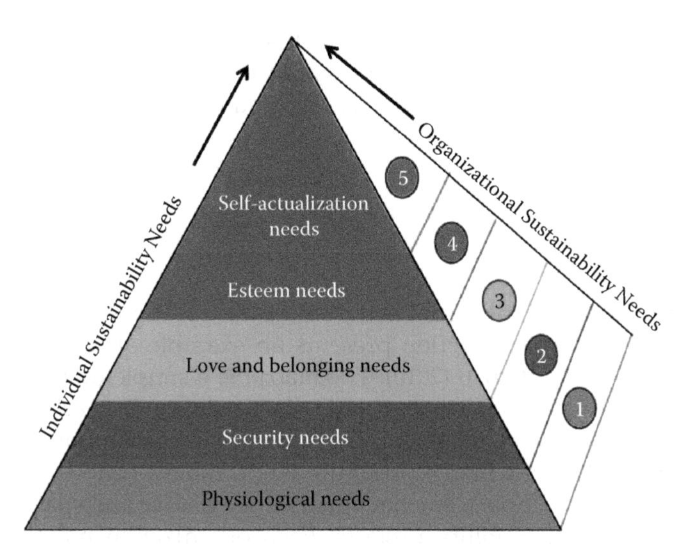

FIGURE 2.1
3-D hierarchy of needs for sustainability goals.

Level 1 of Organizational Needs: This is the organizational need for basic essentials of economic vitality to support the provision of value for stockholders and employees. Can the organization fund projects from cash reserves? *Sustainability applies here.*

Level 2 of Organizational Needs: This is a need for organizational defense. Can the organization feel safe from external attack? Can the organization protect itself cyber attacks or brutal takeover attempts? *Sustainability applies here.*

Level 3 of Organizational Needs: This is the need for an organization to belong to some market alliances. Can the organization be invited to join trade groups? Does the organization have a presence on some world stage? *Sustainability applies here.*

Level 4 of Organizational Needs: This is the level of having market respect and credibility. Is the organization esteemed in some aspect of market, economic, or technology movement? What positive thing is the organization known for? *Sustainability applies here.*

Level 5 of Organizational Needs: This is the level of being classified as a "power" in the industry of reference. Does the nation have a recognized niche in the market? *Sustainability applies here.*

Obviously, where the organization stands in its hierarchy of sustainability goals will determine how it influences its employees (as individuals) to embrace, support, and practice sustainability. How each individual responds to organizational requirements depends on that individual's own level in the hierarchy of needs. We must all recognize the factors that influence sustainability in our strategic planning programs. For an organization to succeed, sustainability must be expressed explicitly as a goal across organizational functions.

Sustainability Matrix

The coupling of technical assets and managerial tools is essential for realizing sustainability. This section presents an example of the *sustainability matrix* introduced by Badiru (2010). The matrix is a simple tool for organizing the relevant factors associated with sustainability. It overlays sustainability awareness factors, technical assets, and managerial tools. Figure 2.2 shows an example of the matrix structure while Table 2.1 shows the generic matrix design. The sample elements illustrate the nature and span of factors associated with sustainability projects. Each organization must assess its own environment and include relevant factors and issues within the context of prevailing sustainability programs. Without a rigorous analytical framework, sustainability will just be in talks rather than deeds. One viable strategy is to build collaborative STEM (Science, Technology, Engineering, and Mathematics) alliances for sustainability projects. The analytical framework of systems engineering provides a tool for this purpose from an interdisciplinary perspective. With this, environmental systems, industrial systems, and societal systems can be sustainably tied together to provide win-win benefits for all. An effective collaborative structure would include researchers and practitioners from a wide variety of disciplines (civil and environmental

FIGURE 2.2
Factors of sustainability.

TABLE 2.1

Generic Template for Sustainability Matrix

	Managerial Environmental Factors				
Technical Factors	**Organizational Behavior**	**Personnel Culture**	**Resource Base**	**Market Influence**	**Share Capital**
Physical infrastructure	Communication modes	Cooperation incentives	Coordination techniques	Building performance	Energy economics
Work design	Technical acquisitions	Work measurement	Project design	Financial implications	Project control
Analytical modeling	Resource combinations	Qualitative risk	Engineering analysis	Value assessment	Forecast models
Scientific limitation	Fuel efficiency	Technical workforce	Contingency planning	Contract administration	Green purchases
Technology constraints	Energy conservation	Training programs	Quantitative risk	Public acceptance	Technology risks

engineering, industrial engineering, mechanical engineering, public health, business, etc.).

Project sustainability is as much a need as the traditional components of project management, which spans planning, organizing, scheduling, and control. Proactive pursuit of the best practices of sustainability can pave the way for project success on a global scale. In addition to people, technology, and process issues, there are project implementation issues. In terms of performance, if we need a better policy, we can develop it. If we need technological advancement, we have capabilities to achieve it. The items that are often beyond reach relate to project life cycle management issues. Project sustainability implies that sustainability exists in all factors related to the project. Thus, we should always focus on project sustainability.

Think "sustainability" in all you do and you are bound to reap the rewards of better resource utilization, operational efficiency, and process effectiveness. Both management and technical issues must be considered in the pursuit of sustainability. People issues must be placed at the nexus of all the considerations of sustainability. Otherwise, sustainability itself cannot be sustained. Many organizations are adept at implementing rapid improvement events (RIE). This chapter recommends a move from mere RIE to sustainable improvement events (SIE).

Social Change for Sustainability

Change is the root of advancement. Sustainability requires change. Our society must be prepared for change to achieve sustainability. Efforts that support sustainability must be instituted into every aspect of everything that the society does. If society is better prepared for change, then positive changes can be achieved. The "pain but no gain" aspects of sustainability can be avoided if proper preparations have been made for societal changes. Sustainability requires an increasingly larger domestic market to preserve precious limited natural resources. The social systems that make up such markets must be carefully coordinated. The socioeconomic impact on sustainability cannot be overlooked.

Social changes are necessary to support sustainability efforts. Social discipline and dedication must be instilled in the society to make sustainability changes possible. The roles of the members of a society in terms of being responsible consumers and producers of consumer products must be outlined. People must be convinced of the importance of the contribution of each individual whether that individual is acting as a consumer or as a producer. Consumers have become so choosy that they no longer will simply accept whatever is offered in the marketplace. In cases where social dictum directs consumers to behave in ways not conducive to sustainability, gradual changes must be facilitated. If necessary, an acquired taste must be developed to like and accept the products of local industry. To facilitate consumer acceptance, the quality of industrial products must be improved to competitive standards. In the past, consumers were expected to make do with the inherent characteristics of products regardless of potential quality and functional limitations. This has changed drastically in recent years. For a product to satisfy the sophisticated taste of the modern consumer, it must exhibit a high level of quality and responsiveness to the needs of the consumer with respect to global expectations. Only high-quality products and services can survive the prevailing market competition and, thus, fuel the enthusiasm for further sustainability efforts. Some of the approaches for preparing a society for sustainability changes are listed below:

- Make changes in small increments
- Highlight the benefits of sustainability development
- Keep citizens informed of the impending changes
- Get citizen groups involved in the decision process
- Promote sustainability change as a transition to a better society
- Allay the fears about potential loss of jobs due to new sustainability programs
- Emphasize the job opportunities to be created from sustainability investments

Addressing the above issues means using a systems view to tackle the various challenges of executing sustainability. As has been discussed in the preceding sections, the concept of sustainability is a complex one. However, with a systems approach, it is possible to delineate some of its most basic and general characteristics. For our sustainability purposes, a system is simply defined as a set of interrelated elements (or subsystems). The elements can be molecules, organisms, machines, machine components, social groups, or even intangible abstract concepts. The relations, interlinks, or "couplings" between the elements may also have very different manifestations (e.g., economic transactions, flows of energy, exchange of materials, causal linkages, control pathways). All physical systems are *open* in the sense that they have exchanges of energy, matter, and information with their environment that are significant for their functioning. Therefore, what the system "does," in its behavior, depends not only on the system itself but also on the factors, elements, or variables coming from the environment of the system. The environment impacts "inputs" onto the system while the system impacts "outputs" onto the environment. This, in essence, is the systems view of sustainability.

Reference

Badiru, A. B. (2010). "The Many Languages of Sustainability," *Industrial Engineer*, Nov., pp. 31–34.

3

Lean and Waste Reduction

Every defect is a treasure, if the company can uncover its cause and work to prevent it across the corporation.

—**Kiichiro Toyoda, founder of Toyota**

Lean is a term that is well known and defined as an elimination of waste in operations through managerial principles. Many principles are included in the Lean concept, but the major thought to remember is effective utilization of resources and time to achieve higher-quality products and ensure customer satisfaction. Remembering back, defects are anything that the customer is unhappy with and is a term utilized in Six Sigma. Six Sigma identifies and eliminates these defects so that the customer in turn is satisfied. Customers are the number one focus and if they are unhappy, they will have no problem going elsewhere, which most likely is a competition for the business. Coupling Lean and Six Sigma will reduce waste and reduce defects. The concept will be called Lean Six Sigma going further.

The most basic concept when discussing waste reduction begins with Kaizen. *Kaizen* is a Japanese concept defined as "taking apart and making better." The concept takes a vast amount of project management techniques to facilitate the process going forward. 5S processes are the most predominant and commonly known for Kaizen events. 5S principles are determined by finding a place for everything and everything in its place

The 5S levels are as follows:

Sort—Identify and eliminate necessary items and dispose of unneeded materials that do not belong in an area. This reduces waste, creates a safer work area, opens space, and helps visualize processes. It is important to sort through the entire area. The removal of items should be discussed with all personnel involved. Items that cannot be removed immediately should be tagged for subsequent removal.

Sweep—Clean the area so that it looks like new and clean it continuously. Sweeping prevents an area from getting dirty in the first place and eliminates further cleaning. A clean workplace indicates high standards of quality and good process controls. Sweeping should eliminate dirt, build pride in work areas, and build value in equipment.

Straighten—Have a place for everything and everything in its place. Arranging all necessary items is the first step. It shows what items are required and what items are not in place. Straightening aids efficiency; items can be found more quickly and employees travel shorter distances. Items that are used together should be kept together. Labels, floor markings, signs, tape, and shadowed outlines can be used to identify materials. Shared items can be kept at a central location to eliminate purchasing more than needed.

Schedule—Assign responsibilities and due dates to actions. Scheduling guides sorting, sweeping, and straightening and prevents regressing to unclean or disorganized conditions. Items are returned where they belong and routine cleaning eliminates the need for special cleaning projects. Scheduling requires checklists and schedules to maintain and improve neatness.

Sustain—Establish ways to ensure maintenance of manufacturing or process improvements. Sustaining maintains discipline. Utilizing proper processes will eventually become routine. Training is key to sustaining the effort and involvement of all parties. Management must mandate the commitment to housekeeping for this process to be successful.

The benefits of 5S include (1) a cleaner and safer workplace; (2) customer satisfaction through better organization; and (3) increased quality, productivity, and effectiveness.

Kai is defined as "to break apart or disassemble so that one can begin to understand." *Zen* is defined as "to improve." This process focuses on improvements objectively by breaking down the processes in a clearly defined and understood manner so that wastes are identified, improvement ideas are created, and wastes are both identified and eliminated. The philosophy includes reducing cycle times and lead times in turn increasing productivity, reducing work-in-process (WIP), reducing defects, increasing capacity, increasing flexibility, and improving layouts through visual management techniques.

Operator cycle times need to be understood to reduce the nonproductive times. Operators should also be cross-functional so that they are able to perform different job functions and the workloads of each function are well balanced. The work performed needs to be not only value-added work but also work that is in demand through customers. WIP should be eliminated to reduce inventory. Inventory should be seen simply as money waiting in process and should be reduced as much as possible. WIP can be reduced by reducing setup times, transporting smaller quantities of batch outputs, and line balancing. Bottlenecks should be removed by finding non-value-added tasks and removing the excess time spent by both machinery and humans. Flexible layouts promote efficiency in the 5Ms, which are defined below:

The 5Ms demonstrated above in Figure 3.1 lead to 7 Wastes in Figure 3.2, which are demonstrated below. Sometimes an 8th Waste is added in and an abbreviation of DOWNTIME is associated with the acronym. It is defined below:

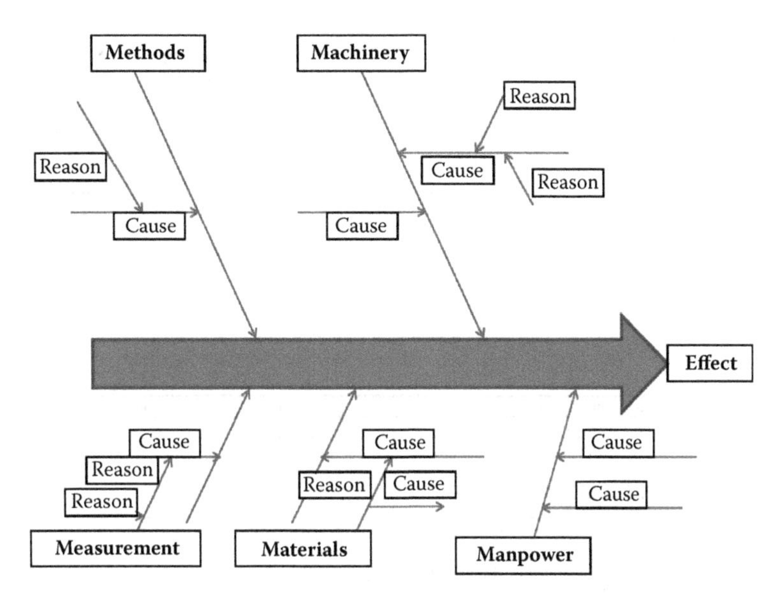

FIGURE 3.1
Cause-and-effect diagram demonstrating the 5Ms.

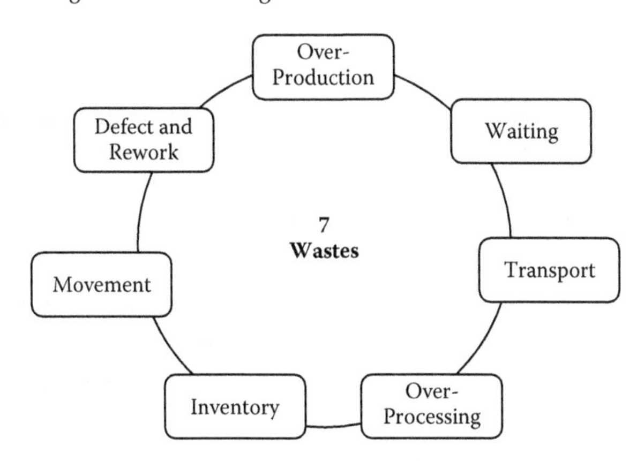

FIGURE 3.2
Illustration of 7 Wastes.

- Defect/correction
- Overproduction
- Waiting
- Not utilizing employee talents
- Transportation/material movement
- Inventory
- Motion
- Excessive processing

Root Cause Analysis			
Root Cause Analysis: Finding causes of problems while implementing action plans in order to eliminate future instances of occurrence and maximize continuous improvement			Date:
Issue Date:	RCA Initiated by:		
Inspector:			
1. Define the problem	Describe the issue in detail. What was defect, how many, how often, etc.?		

DMAIC Methodology with 10 Steps		Date completed
D	1 Define the problem	
	2. Process Mapping	
M	3. Data Gathering	
	4. Cause/Effect Analysis (Seeking Root Cause) Utilize 5Ms	
A	5. Verifying root cause with data driven results	
	6. Solutions & Continuous Improvement (Include Costs and Benefits)	
I	7. Test Implementation Plan by Piloting	
	8. Implementation	
C	9. Control/Monitoring Plan	
	10. Documentation of Lessons Learned	

FIGURE 3.3
Problem definition.

The primary technique for reducing waste and defects utilizing Lean Six Sigma techniques is demonstrated below in order of operations:

1. Define the problem—See Figure 3.3.
2. Process mapping—See Figure 3.4.
3. Data gathering—Gather data on any process with defects and issues. Utilize voice of the customer to find what data is needed.
4. Cause/effect analysis (seeking root cause) utilizing 5Ms—See Figure 3.5.
5. Verifying root cause with data-driven results—Ask why five times to ensure the proper root cause is found. Do not Band-Aid problems; instead eliminate the cause of their occurrence.
6. Solutions and continuous improvement plans—See Figure 3.6. Include costs and benefits, if there are financial paybacks, they should be included—See Figure 3.7.
7. Test implementation plan by piloting. Pilot plans can include actual trials or mock trials with details information.
8. Implement continuous improvement ideas.
9. Control/monitoring plan—See Figure 3.8.
10. Documentation of lessons learned—See Figure 3.9.

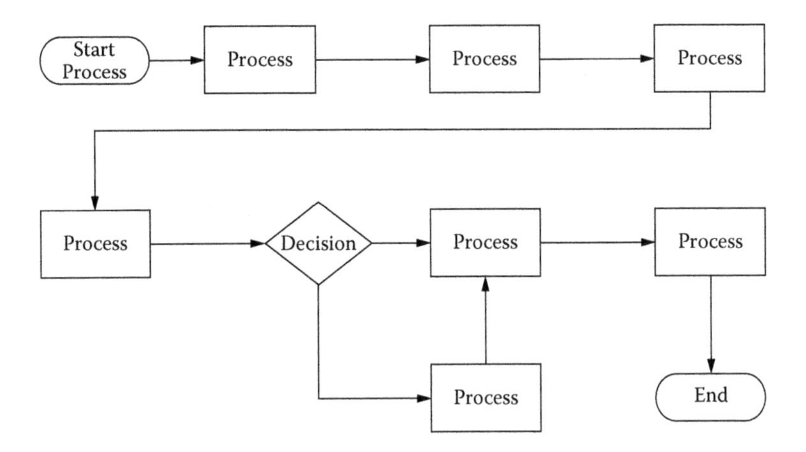

FIGURE 3.4
Perform process mapping.

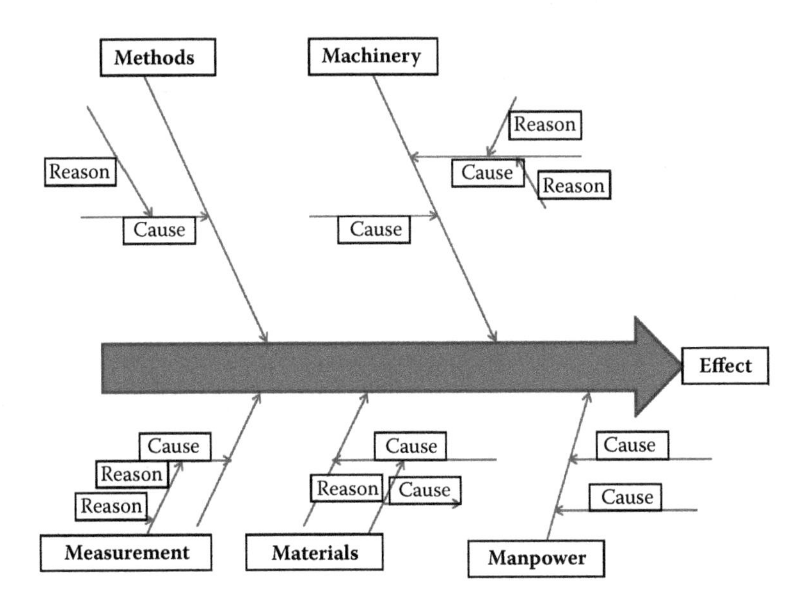

FIGURE 3.5
Cause/effect analysis.

Now that the process is laid out in terms of making proper improvements, the sustainability portion must be realized. The control plan during this phase is crucial. The control plan not only must be documented but also should be a living document that is followed structurally. The accountability portion for this phase is a key portion in having lasting results. The more specific the plan is, the better off the implementation of it will be. The plans must also be attainable, or the plan will fail.

Top Level
Next Level
Final Level
Idea
Idea
Idea
Idea
Idea
Idea

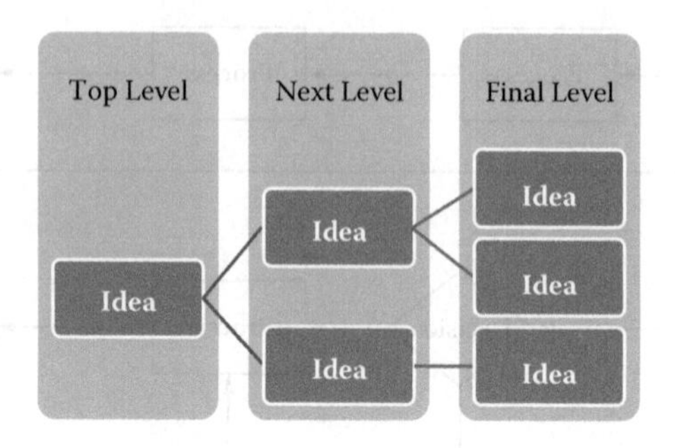

FIGURE 3.6
Recommendations and continuous improvement plans.

Cost Benefit			
Annualized cost of issue		1	$0.00
Percent defect reduction		2	0%
Cost of proposed solution		3	$0.00
Financial Payback for Year 1 (1 × 2 – 3)			$0.00

FIGURE 3.7
Cost-benefit analysis.

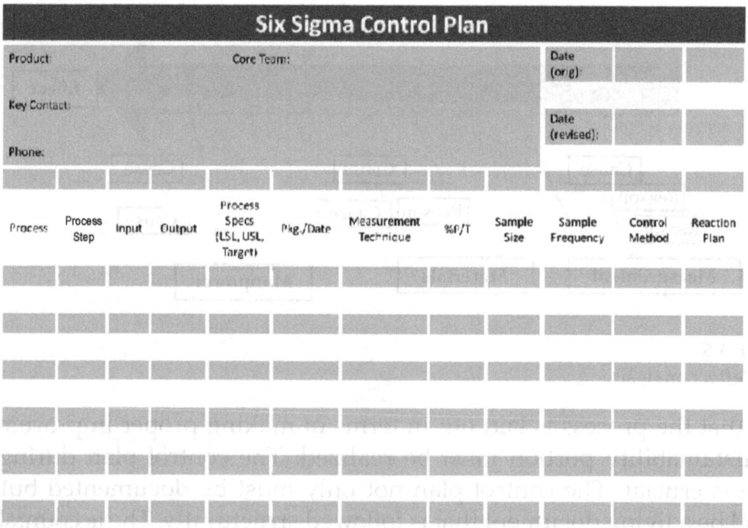

Six Sigma Control Plan
Product:
Core Team:
Date (orig):
Key Contact:
Date (revised):
Phone:
Process
Process Step
Input
Output
Process Specs (LSL, USL, Target)
Pkg./Date
Measurement Technique
%P/T
Sample Size
Sample Frequency
Control Method
Reaction Plan

FIGURE 3.8
Control plan.

Name: Date: Area:

One Point Lesson

Topic

Topic Goes Here

Objective:

Be SMART: Specific, Measureable, Attainable, Relevant, and Time-Bound

Process:

Explain Process Here.

Example:

TITLE

Process

Date/ product number	Reasons

FIGURE 3.9
Documentation of lessons learned.

Lean can also involve some statistical tools. The tools demonstrate the efficiencies and labor balancing. The main statistical tools are described below.

First pass yield (FPY) indicates the number of good outputs from a first pass at a process or step. The formula is as follows:

$$\text{FPY} = (\text{\# accepted})/(\text{\# processed})$$

The formula for the first pass yield ratio is % FPY = [(# accepted)/(# processed)] × 100. This number does not include reworked product that was previously rejected.

Rolled throughput yield (RTY) covers an entire process. If a process involves three activities with FPYs of 0.90, 0.94, and 0.97, the RTY would be 0.90 × 0.94 × 0.97 = 0.82. The %RTY = 0.82 × 100 = 82%.

Value-added time (VAT) is % VAT = (sum of activity times)/(lead time) × 100. When the sum of activity times equals lead time, the value-added time is 100%. For most processes, % VAT = 5 to 25%. If the sum of activity times equals the lead time, the time value is not acceptable and activity times should be reduced.

Takt time is a kaizen tool used in the order-taking phase. *Takt* is a German word for "pace." Takt time is defined as time per unit. This is the operational measurement to keep production on track. To calculate Takt time, the formula is time available/production required. Thus, if a required production is 100 units per day and 240 minutes are available, the Takt time = 240/100 or 2.4 minutes to keep the process on track. Individual cycle times should be balanced to the pace of Takt time. To determine the number of employees required, the formula is (labor time/unit)/Takt time. Takt in this case is time per unit. Takt requires visual controls and helps reduce accidents and injuries in the workplace. Monitoring inventory and production WIP will reduce waste or muda. *Muda* is a Japanese term for waste where waste is defined as any activity that consumes some type of resource but is non-value-added for the customer. The customer is not willing to pay for this resource because it is not benefiting them. Types of muda include scrap, rework, defects, mistakes, and excess transport, handling, or movement.

The Lean House is a common methodology for understanding Lean and waste reduction. The house looks as shown in Figure 3.10.

Mistake proofing is a subject of its own when brought into the Lean Six Sigma methodology. This term is often called Poka-Yoke, also known as another initiative to improve production systems. The methodology eliminates product defects before they occur by simply installing processes to prevent the mistakes from happening in the first place. These mistakes that happen are due to human nature and can normally not be eliminated by simple training or SOPs. These steps to eliminate the defect will prevent the next step in the process from occurring if a defect is found. Normally there is some type of alert that will show there is a mistake and will prevent the process from going forward. An example of a Poka-Yoke would be a simple check weigher that would kick off a package of food if it were not the correct weight.

Poka-Yokes often also encompass a concept called zero quality control (ZQC). This does not mean a reduction in defects, but instead complete elimination of defects, also known as zero defects. ZQC was another concept led by the Japanese that leads to low-inventory production. The reason the inventory is so low is due to not needing excess inventory due to having to replace defective parts less often. ZQC also focuses on quality control and data versus blaming humans for mistakes. The methodology was developed by Shigeo Shingo, who knew it was human nature to make common mistakes and did not feel people should be reprimanded for them. Shingo said, "Punishment makes people feel bad; it does not eliminate defects."

This concept is important because it focuses on the customers and realizes that defects are costly; therefore eliminating defects saves money. Many companies rework product to save money but do not make the effort to eliminate the problem in the first place. This process will eliminate rework by eliminating any defects from happening in the first place.

The first cycle of the ZQC system is the Plan-Do-Check cycle, also known as PDC, shown in Figure 3.11.

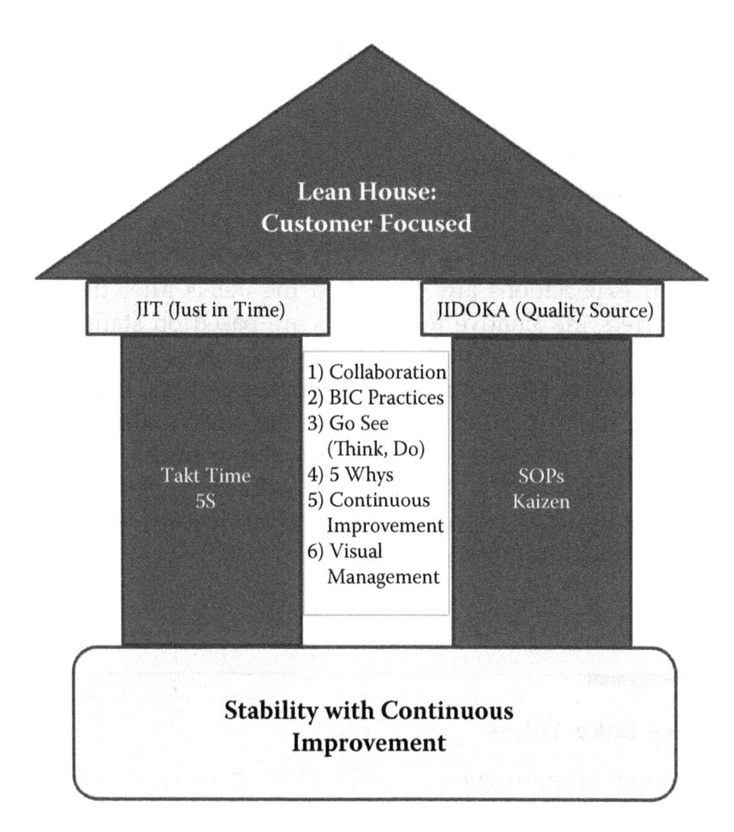

FIGURE 3.10
Illustration of Lean House.

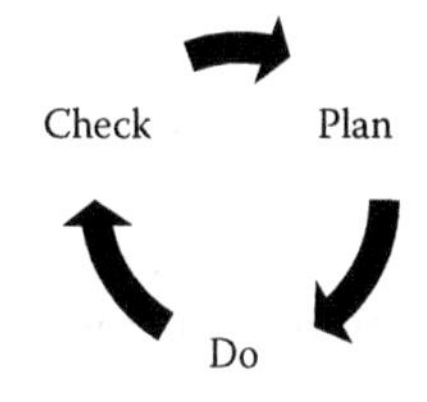

FIGURE 3.11
Plan-Do-Check illustration.

This is a traditional cycle where processes and conditions are planned out, the planned actions are performed in the Do phase, and finally quality control checks are performed in the Check phase. This method catches mistakes and also provides feedback during the Check phase. The checks in this phase also account for 100% inspection; therefore all parts or processes are looked upon, indicating no defects.

There are three main types of checks or inspections that are popular:

- Judgment inspections
- Informative inspections
- Source inspections

Judgment Inspections are those that are done normally by humans based on what their expectations are. They find the defect after the defect has already occurred. Informative inspections are based on statistical quality control (SQC), checks on each product, and self-checks. These inspections help reduce defects but do not eliminate them completely. Finally the source inspections are the inspections that reduce the defects completely. Source inspections discover the mistakes before processing and then provide feedback and corrective actions so that the process has zero defects. The source inspections require 100% inspection. The feedback loop is also very quick so that there is minimal waiting time.

How to Use Poka-Yokes

Poka-Yokes use two approaches:

- Control systems
- Warning systems

Control systems stop the equipment when a defect or unexpected event occurs. This prevents the next step in the process from occurring so that the complete process is not performed. Warning systems signal operators to stop the process or address the issue at the time. Obviously the first of the two prevents all defects and has a more ZQC methodology because an operator could be distracted or not have time to address the problem. Control systems often also use lights or sounds to bring attention to the problem; that way, the feedback loop again is very minimal.

The methods for using Poka-Yoke systems are as follows:

- Contact methods
- Fixed-value methods
- Motion-step methods

Contact methods are simple methods that detect whether products are making physical or energy contact with a sensing device. Some of these are commonly known as limit switches where the switches are connected to cylinders and pressed in when the product is in place. If a screw is left out, the product does not release to the next process. Other examples of contact methods are guide pins.

Fixed-value methods are normally associated with a particular number of parts to be attached to a product or a fixed number of repeated operations occurring at a particular process. Fixed-value methods utilize devices as counting mechanisms. The fixed-value methods may also use limit switches or different types of measurement techniques.

Finally, the motion-step method senses if a motion or step in the process has occurred in a particular amount of time. It also detects sequencing by utilizing tools such as photoelectric switches, timers, or bar-code readers.

The conclusion of Poka-Yokes is to use the methodology as mistake proofing for ZQC to eliminate all defects, not just some. The types of Poka-Yokes do not have to be complex or expensive, just well thought out to prevent human mistakes or accidents.

The Poka-Yoke discussion branches into the correct location discussion. This technique places design and production operations in the correct order to satisfy customer demand. The concept is to increase throughput of machines to ensure that the production is performed at the proper time and place. Centralization of areas helps final assemblers, but the most common practice to be effective is to unearth an effective flow. U-shaped flows normally prevent bottlenecks. Value-stream mapping is a key component during this time to establish that all steps occurring are adding value. Keep in mind that value-added activities are any activities that the customer is willing to pay for. Another note to remember is to not only have a smart and efficient technique but also only produce goods that the customer is demanding to eliminate excess inventory.

This technique is called the pull technique. Pull is the practice of not producing any goods upstream if the downstream customer does not need them. The reason this is a difficult technique is because once an efficient method is found to produce a good, the mass production begins. The operations forget if the goods are actually needed or not and begin thinking only of throughput. Even though co-manufacturers seem like a bad idea for many employers, they sometimes come in handy when a small amount of a versatile product is needed.

Push systems, on the other hand, are not effective due to predictions of customer demands.

Lean systems show the pull system utilizing machinery for 90% of requirements and limit downtime to 10% for changeovers and maintenance. This does not mean preventative maintenance should not be performed, but only that the maintenance time is reduced to 10%. Kanbans are a key factor during this Lean system in order to use a visual indicator that another part or process is required. This also prevents excess parts from being made or excess processes being performed.

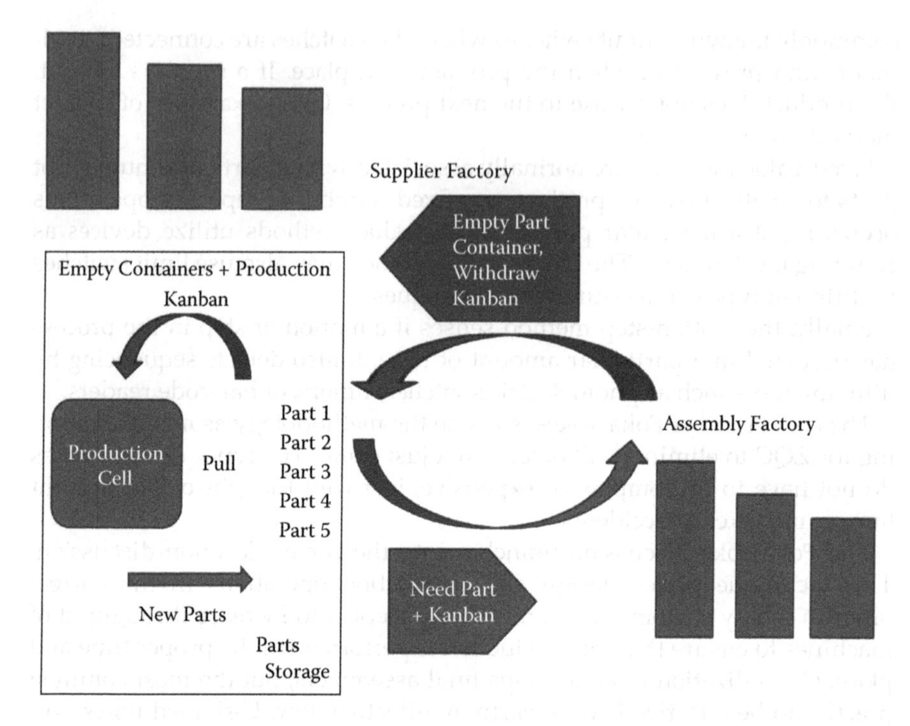

FIGURE 3.12
Pull flow system.

Heinjunka is the leveling of production and scheduling based on volume and product mix. Instead of building products according to the flow of customer orders, this technique levels the total volume of orders over a specific time so that uniform batches of different product mixes are made daily. The result is a predictable matrix of product types and volumes. For heinjunka to succeed, changeovers must be managed easily. Changeovers must be as minimally invasive as possible to prevent time wasted because of product mix. Another key to heinjunka is making sure that products are needed by customers. A product should not be included in a mix simply to produce inventory if it is not demanded by customers. Long changeovers should be investigated to determine the reason and devise a method to shorten them. A pull flow diagram is shown in Figure 3.12.

The Lean action plan is simply drawn out in five steps:

1. Getting started—Plan out the appropriate steps. This will take 1 to 6 months
2. Create the new organization and restructure. This will take 6 to 24 months
3. Implement Lean techniques and systems and continually improve. This will take 2 to 4 years

4. Complete the transformation. This will take up to 5 years.
5. Do the entire process again to have another continuous improvement project and sustain the results.

References

Kovach, T. (2012). *Statistical Techniques for Project Control*. Taylor & Francis, New York.

The Productivity Press Development Team. (2010). *Mistake-Proofing for Operators: The ZQC System*. Taylor & Francis, New York.

4

Education and Sustainability

Education is learning what you didn't even know you didn't know.

Daniel J. Boorstin

Education is the most powerful tool that can be used toward sustainability. Education personally, economically, or socially is important to improving the bottom line of a business or individual. Sustainability practices save money through awareness and communication while reducing environmental predicaments. Engagement of individuals whether personally or professionally will help engage people to understand their direct responsibility and the effect they have on the environment. When businesses are going through tough times, the number one thing to do is cut costs. Many think cutting jobs is the first answer but should think first about how to be more sustainable. Innovation mixed with business processes can change mindsets of people and businesses and reduce incremental costs. Education plays a vital role with these processes because it changes people's ways of thinking. The methodology is to encourage people to perform certain functions not only while at work but also in their everyday lives. This is a life change, not a flavor of the month. Without making the changes, our entire environment will suffer consequences.

Education can be the easiest part of sustainability. Education includes engagement, motivation, teaching, and changes of everyday processes, all leading to continual improvements toward our day-to-day lives. Forming teams for sustainability is a great initiative to begin educating others. As a team, each person puts forward his or her own understanding and education of sustainability and participates in a collaborative effort toward the initiative. The beginning process of this is to form a proper team. The team must be cross-functional, knowing vast areas of the entire environment. A project charter is the next step for this team so there is an executive summary and understanding of what is to be accomplished. Once the project charter is developed, a timeline is imperative so the focus does not diminish and the priority remains high. Senior management should be given a presentation on the group's objectives and goals along with permission for resources for the effort. A definition for sustainability should be identified next. This definition can vary slightly but should include some of the following terms or ideas:

> Sustainability is the human preservation of the environment, whether personally, socially, or economically through responsible management of resources, education, and continuous process improvement. Being sustainable corporate citizens will increase sales and profitability by reducing costs and having a competitive advantage.

Once sustainability is defined, the next steps are the action steps. Communication is a must on what sustainability means to the remaining core groups, which is then cascaded down to everyone involved in the corporation or area. An action plan should come next on how to be sustainable and the steps involved. Each business is different, but commonalities are energy reduction utilizing automated thermostats, turning off computers, utilizing less water, and decreasing gas hot water heaters.

The work log in Table 4.1 can be utilized to ensure progress.

The following statement was written by the participants in Business Sector Team calls of the U.S. Partnership for Education for Sustainable Development:

> To meet the immense challenges of the present and the future, it is important that all undergraduate and graduate college students learn about our environmental and social sustainability challenges and be provided with learning opportunities that engage them in solutions to these challenges. We live in a unique time, where the decisions of this generation may very well dictate the health of the planet for this and future generations. The impacts of these decisions will affect the quality of life across the globe. All students need to learn, through an interdisciplinary approach, not only the specifics of our sustainability challenges and the possible solutions, but also the interpersonal skills, the systems thinking skills, and the change agent skills to effectively help to create a more sustainable future. We are looking for these sustainability educated students as future business people, as employees, as consumers, innovators, government leaders and investors. We would like to see this be a requirement for all students.

Stimulus money is increasingly rising for education toward green jobs. The federal economic stimulus plan allocated more than $70 billion in direct spending, tax breaks, and loan guarantees for the energy in the nation, mostly the green energy.

Strategic Role of Sustainability Education

Whoever owns the knowledge controls the power. Education should play a vital role in sustainability programs. In these days of knowledge economy, adequate education is needed to succeed in any workplace whether it is a

TABLE 4.1
Work Log Table

Department	Project	Action	Follow Up	Completion	Annual Savings	Responsible	Due Date	Cross Functional Team	Potential Failure Modes	Notes
Line 1	Water reduction	Turn off water during breaks		● 100%	$10,000	Fred	16-Jan		Employees forget	Work order written
		Turn off water after tool is washed		◆ 25%	$1,000	George	27-Jan			
		Use cold water versus hot when QA has approved		◆ 0%	$7,500	Jenna	18-Jan			Ensure QA has approved
		Turn down hot water heater		◆ 0%	$1,000	Tina	24-Jan		Preventative maintenance	
		Fix leaking faucet in bathroom 3 by break room		● 100%	$1,550	Tina	27-Jan			Work order written
		Utilize hose reels to ensure hoses are put away and not left on		● 0%	$1,000	Tina	31-Jan			Need to try to find a cheaper solution

hamburger stand, an administrative office, or a manufacturing plant. Even in some industrialized nations, a large percentage of the adult population in neglected communities is functionally illiterate and less likely to embrace sustainability. It used to be that the children of these poverty-stricken communities were needed to drive the wheels of manual labor in local factories. But modern industries, with increasing push for automation, do not need much of the services of the low-grade workers. Knowledge workers are the norm in the present economy. Education geared toward the new sustainability direction will, thus, be needed to participate actively in the society. Poor parents always proclaim that they don't want their children to end up where they did. Yet, only limited educational efforts are made to ensure that they don't.

University–Industry Sustainability Partnership

Knowledge is an everlasting capital. The establishment of a formal process for the interface of institutions of higher learning and industry can be one of the capitals for achieving sustainability. Universities have unique capabilities that can be aligned with industry capabilities to produce symbiotic working relationships. Private industrial research projects must complement public industrial research ventures for the sake of advancing sustainability.

Academic institutions have a unique capability to generate, learn, and transfer technology to industry. The quest for knowledge in academia can fuel the search for innovative solutions to specific sustainability problems. Cooperating industry is a fertile ground for developing prototypes of new innovation. Industrial settings are good avenues for practical implementation of technology. Industry-based implementation of university-developed sustainability technology can serve as the impetus for further efforts to develop new technology. Technologies that are developed within the academic community mainly for research purposes often fade into oblivion because of the lack of formal and coordinated mechanism for practical implementation. The potentials of these technologies go untapped for several reasons including the following:

- The developer does not know which industry may need the technology.
- Industry is not aware of the technology available in academic institutions.
- There is no coordinated mechanism for technical interface between industry and university groups.

Universities interested in technology development are often hampered by the lack of adequate resources for research and training activities. Industry can help in this regard by providing direct support for industrial groups to address specific industrial problems related to sustainability. The universities also need real problems to work on as projects or case studies. Industry can provide these under a cooperative arrangement.

The respective needs and capabilities of universities and industrial establishments can be integrated symbiotically to provide benefits for each group. University courses offered at convenient times for industry employees can create opportunity for university–industry interaction. Class projects for industry employees can be designed to address real-life sustainability problems. This will help industry employees to have focused and rewarding projects. Class projects developed in the academic environment can be successfully implemented in actual work environments to provide tangible benefits. With a mutually cooperative interaction, new sustainability developments in industry can be brought to the attention of academia while new academic research developments can be tested in industrial settings.

Sustainability Clearinghouse

Academic institutions can serve as convenient locations for technology clearinghouses. Such clearinghouses can be organized to provide up-to-date information for sustainability activities. Specific industrial problems can be studied at the clearinghouse. The clearinghouse can serve as a repository for information on various technology tools for sustainability. Industry would participate in the clearinghouse through the donation of equipment, funds, and personnel time. The services provided by a clearinghouse could include a combination of the following:

- Provide consulting services on technology to industry
- Conduct on-site short courses with practical projects for industry
- Serve as a technology library for general information
- Facilitate technology transfer by helping industry identify which technology is appropriate for which problems
- Provide technology management guidelines that will enable industry to successfully implement new technology in existing operations
- Expand training opportunities for engineering students and working engineers

Center of Excellence for Sustainability

As the interest in sustainability spreads, there will be a need to differentiate capabilities among the various players. In its oversight roles, the government can sponsor cooperative interactions between academia and industry by providing broad-based funding mechanisms. As an example, the United States National Science Foundation (NSF) has a program to provide funds for Industry/University Cooperative Research Centers. This sort of partnership can be leveraged for the pursuit of sustainability programs. The leaders of a nation pursuing sustainability should actively support cooperative efforts between universities and industry. The establishment of centers of excellence for pursuing sustainability-related research is one approach to creating an atmosphere that is conducive for industry-university interaction. Several states in the United States have used this approach to address specific development needs.

The Role of Women in Achieving Sustainability

Even though women may not generally be the head of the household, they are, nonetheless, leaders of the home. In this leadership role, women can be key players in ensuring sustainability regardless of whichever sustainability definition is in operation. Women play important roles in national development and should lead societal efforts in sustainability. Women have significant capabilities in terms of political, economic, and industrial roles in a nation. In developing nations particularly, women have played important economic roles as traders for a long time. The economic power garnered from their entrepreneurial activities should be transformed to decision-making power to support sustainability programs. In very democratic nations, the political power of women has been more pronounced primarily through voting rights. These rights should be directed at policy-making endeavors that can facilitate sustainability. Women's groups have emerged in some developing nations to better utilize their collective powers. Many of the efforts of these groups have been directed at development. For example, the National Council on Women and Development has been very active politically in Ghana. In Nigeria, the Rural Women Development program, spearheaded by the first lady of the nation, is directed at improving the economic impact of rural women. Small-scale industries owned by women have been one of the tangible outputs of these women's movements. Other examples abound in developed and developing nations around the world. It is the view of the authors that there should be a global coalition of women for sustainability, whereby an educational platform will be a primary strategy.

Agriculture and Sustainability

Agriculture is one avenue through which the impacts and benefits of sustainability can be readily noticed. A hungry society cannot be an industrially productive society and cannot be focused on sustainability initiatives. If the masses are fed and feel content, then they will have room to consider and embrace sustainability. It is generally believed that an underdeveloped economy is characterized by an agricultural base. Based on this erroneous belief, several developing nations have abandoned their previously solid agricultural base in favor of "unsustainable" industrialization. The fact is that a strong agricultural base is needed to complement industrialization programs. Agriculture, itself, is a good target for modernization and industrialization in ways that complement and support sustainability goals. If industrialization does not yield immediate benefits, the society will be exposed to the double jeopardy of hunger and material deprivation. Once abandoned, agriculture is a difficult process to recoup and non-sustainable practices will creep in. Since agricultural processes take several decades to perfect, revitalization of abandoned agriculture may require several decades. Agriculture should play a major role in the foundation for sustainable industrial development. The agricultural sector can serve as a viable market for a developed industry.

Evolution of Efficient Agriculture

It is interesting to note how the agricultural revolution led to the industrial revolution in the past. Human history indicates that humans started out as nomad hunters and gatherers, drifting to wherever food could be found. About 12,000 years ago, humans learned to domesticate both plants and animals. This agricultural breakthrough allowed humans to become settlers, thereby spending less time in search of food. More time was, thus, available for pursuing innovative activities, which led to discoveries of better ways of planting and raising animals for food. That initial agricultural discovery eventually paved the way for the agricultural revolution. During the agricultural revolution, mechanical devices, techniques, and storage mechanisms were developed to aid the process of agriculture. These inventions made it possible for more food to be produced by fewer people. The abundance of food meant that more members of the community could spend that time for other pursuits rather than the customary labor-intensive agriculture. Naturally, these other pursuits involved the development and improvement of the tools of agriculture. The extra free time brought on by more efficient agriculture was, thus, used to bring about minor technological improvements in agricultural implements. These more advanced agricultural tools led to even more efficient agriculture. The transformation from the digging stick to the metal hoe is a good example of the raw technological innovation of that time. Sustainability was possible then, and it should be possible now.

Emergence of Cities

With each technological advance, less time was required for agriculture, thereby, permitting more time for further technological advancements. The advancements in agriculture slowly led to more stable settlement patterns. These patterns led to the emergence of towns and cities. With central settlements away from farmlands, there developed a need for transforming agricultural technology to domicile technology that would support the new city life. The transformed technology was later turned to other productive uses which eventually led to the emergence of the industrial revolution. To this day, the intertwined relationships between agriculture and industry can still be seen, although one would have to look harder and closer to see them from the standpoint of sustainability.

Human Resources for Sustainability

Technical, administrative, and service manpower will be needed to support sustainability programs. People make development possible. People must sustain sustainability activities. No matter how technically capable a machine may be, people will still be required to operate or maintain it. Soon after World War II, it was generally believed that physical capital formation was a sufficient basis for development. That view was probably justified at that time because of the role that machinery played during the war. It was not obvious then that machines without a trained and skillful workforce did not constitute a solid basis for development. It has now been realized that human capital is as crucial to development and sustainability as physical capital is. The investment in human resource development through education must be given a high priority in the overall sustainability strategy. Some of the important aspects of manpower supply analysis for industrial development include the following:

- Level of skills required
- Mobility of the manpower

- The nature and type of skills required
- Retention strategies to reduce brain drain
- Potential for coexistence of people and technology
- Continuing education to facilitate adaptability to technology changes

The Role of Technology in Sustainability

Technology can facilitate sustainability. But technology must be managed properly to play an effective role. There is a multitude of new technologies that has emerged in recent years. Hardware and software technologies are playing more and more roles in sustainability programs. But much more remains to be done in educational programs to develop new technologies specifically for sustainability. It is important to consider the peculiar characteristics of a new technology before establishing adoption and implementation strategies for sustainability. The justification for the adoption of a new technology should be a combination of several factors rather than a single characteristic of the technology. The important characteristics to consider include productivity improvement, improved product quality, reduction in production cost, flexibility, reliability, and safety. An integrated evaluation must be performed to ensure that a proposed technology is justified both economically and technically. The scope and goals of the proposed technology must be established right from the beginning of a sustainability project. This entails the comparison of industry objectives with the overall national goals in the areas discussed below.

Market target: This should identify the customers of the proposed technology. It should also address items such as market cost of the proposed product, assessment of competition, and market share.

Growth potential: This should address short-range expectations, long-range expectations, future competitiveness, future capability, and prevailing size and strength of the competition.

Contributions to sustainability goals: Any prospective technology must be evaluated in terms of direct and indirect benefits to be generated by the technology. These may include product price versus value, increase in international trade, improved standard of living, cleaner environment, safer workplace, and improved productivity.

Profitability: An analysis of how the technology will contribute to profitability should consider past performance of the technology, incremental benefits of the new technology versus conventional technology, and value added by the new technology.

Capital investment: Comprehensive economic analysis should play a significant role in the technology assessment process. This may cover an evaluation of fixed and sunk costs, cost of obsolescence, maintenance requirements, recurring costs, installation cost, space requirement cost, capital substitution potentials, return on investment, tax implications, cost of capital, and other concurrent projects.

Skill and resource requirements: The utilization of resources (manpower and equipment) in the pre-technology and post-technology phases of industrialization should be assessed. This may be based on material input/output flows, high value of equipment versus productivity improvement, required inputs for the technology, expected output of the technology, and utilization of technical and nontechnical personnel.

Risk exposure: Uncertainty is a reality in technology adoption efforts. Uncertainty will need to be assessed for the initial investment, return on investment, payback period, public reactions, environmental impact, and volatility of the technology.

National sustainability improvement: An analysis of how the technology may contribute to national sustainability goals may be verified by studying industry throughput, efficiency of production processes, utilization of raw materials, equipment maintenance, absenteeism, learning rate, and design-to-production cycle.

DEJI Model for Sustainability Assessment

In terms of assessing technology for sustainability implementation, one approach that can be used is the Design, Evaluate, Justify, and Integrate (DEJI) model. The model is unique among process improvement tools and techniques because it explicitly calls for a justification of the technology within the process improvement cycle. This is important for the purpose of determining when a technology should be terminated even after going into production. If the program is justified, it must then be integrated and "accepted" within the ongoing sustainability program of the enterprise. Figure 4.1 illustrates the application of the DEJI model to sustainability technology assessment. The model can be applied across the spectrum of the following elements of an organization:

1. People
2. Process
3. Technology

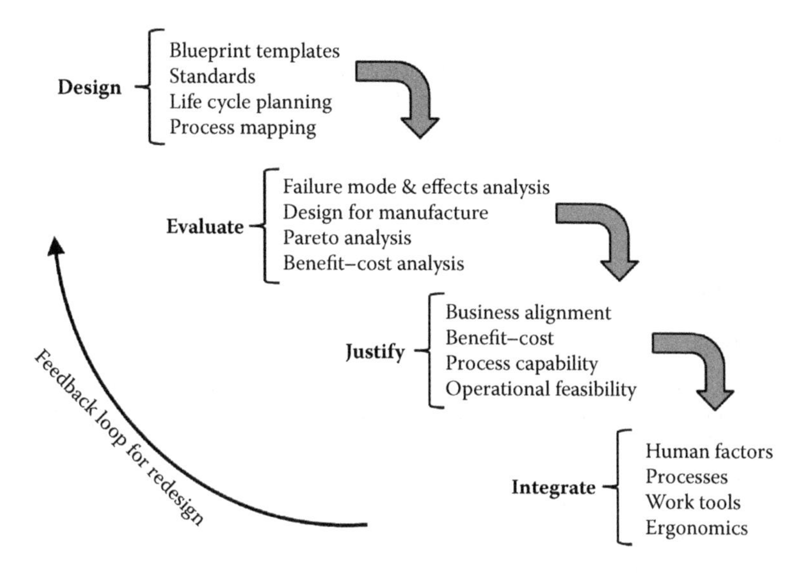

FIGURE 4.1
DEJI model for sustainability technology assessment.

Foundation for Sustaining Sustainability

Sustainability built upon a solid foundation can hardly fail. The major ingredients for durable sustainable industrial, economic, and technological developments are electrical power, water, transportation, and communication facilities. These items should have priority in major sustainability programs.

Primary Amenities:

- Reliable power supply
- Consistent water supply
- Good transportation system
- Efficient communication system

Supporting Amenities:

- Housing
- Education
- Health care

The provision of adequate health care facilities is particularly essential to building a strong industrial base. A healthy society is a productive society;

a sick society will be a destitute society and more susceptible to infringing on sustainability. Diseases that often ravage impoverished nations can curtail the productive capabilities of the citizens. The destructive effects of many of these diseases can be stemmed by prompt access to basic health care services.

References

Badiru, A. B. (2010). "Half-Life of Learning Curves for Information Technology Project Management," *International Journal of IT Project Management*, 1(3), pp. 28–45.

The U.S. Partnership. Accessed from http://www.uspartnership.org/main/view_archive/1.

Reuteman, R. (2011). "Hype Aside, 'Green Jobs' Are for Real," CNBC. Accessed from http://www.cnbc.com/id/43249813.

The National Environmental Education Foundation (NEEF). (2009). *The Engaged Organization: Corporate Employee Environmental Education Survey and Case Study Findings*. Accessed from http://www.neefusa.org/BusinessEnv/EngagedOrganization_03182009.pdf.

5

Six Sigma for Sustainability

A bad system will defeat a good person every time.

—W.E. Deming

Six Sigma is best defined as a business process improvement approach that seeks to find and eliminate causes of defects and errors, reduce cycle times, reduce costs of operations, improve productivity, meet customer expectations, achieve higher asset utilization, and improve return on investment (ROI). Six Sigma deals with producing data-driven results through management support of the initiatives. Six Sigma pertains to sustainability because without the actual data, decisions would be made on trial and error. Sustainable environments require having actual data to back up decisions so that methods are used to have improvements for future generations. The basic methodology of Six Sigma includes a five-step method approach that consists of the following:

Define: Initiate the project, describe the specific problem, identify the project's goals and scope, and define key customers and their Critical to Quality (CTQ) attributes.

Measure: Understand the data and processes with a view to specifications needed for meeting customer requirements, develop and evaluate measurement systems, and measure current process performance.

Analyze: Identify potential cause of problems, analyze current processes, identify relationships between inputs, processes, and outputs, and carry out data analysis.

Improve: Generate solutions based on root causes and data-driven analysis while implementing effective measures.

Control: Finalize control systems and verify long-term capabilities for sustainable and long-term success.

The goal for Six Sigma is to strive for perfection by reducing variation and meeting customer demands. The customer is known to make specifications for processes. Statistically speaking, Six Sigma is a process that produces 3.4 defects per million opportunities. A defect is defined as any event that is outside of the customer's specifications. The opportunities are considered any of the total number of chances for a defect to occur. Table 5.1 explains the defects per million opportunities and sigma levels.

TABLE 5.1

Defects per Million Opportunities and Sigma Levels

Sigma Spread	DPMO	Percent Defective	Percent Yield	Short Term Cpk	Long Term Cpk
1	691,462.00	69%	31%	0.33	–0.17
2	308,538.00	31%	69%	0.67	0.17
3	66,807.00	7%	93.30%	1	0.5
4	6,210.00	0.62%	99.38%	1.33	0.83
5	233	0.02%	99.98%	1.67	1.17
6	3.4	0%	100%	2	1.5

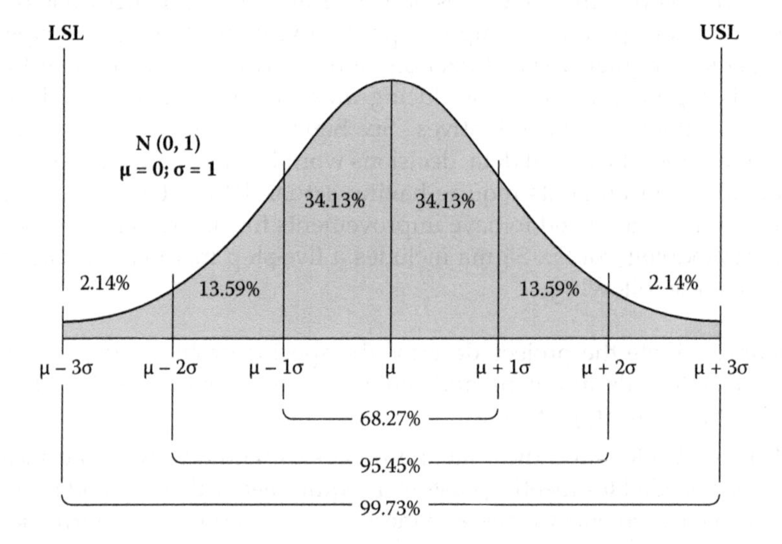

FIGURE 5.1
Areas under the normal curve.

The normal distribution that underlies the statistical models of the Six Sigma model is shown in Figure 5.1.

The Greek letter σ (sigma) marks the distance on the horizontal axis between the mean μ and the curve inflection point. The greater the distance, the greater is the spread of values encountered. The figure shows a mean of 0 and a standard deviation of 1, that is, $\mu = 0$ and $\sigma = 1$. The plot also illustrates the areas under the normal curve within different ranges around the mean. The upper and lower specification limits (USL and LSL) are ±3 σ from the mean or within a six-sigma spread. Because of the properties of the normal distribution, values lying as far away as ±6 σ from the mean are rare because most data points (99.73%) are within ±3 σ from the mean except for processes that are seriously out of control.

Six Sigma allows no more than 3.4 defects per million parts manufactured or 3.4 errors per million activities in a service operation. To appreciate the effect of Six Sigma, consider a process that is 99% perfect (10,000 defects per million parts). Six Sigma requires the process to be 99.99966% perfect to produce only 3.4 defects per million, that is, 3.4/1,000,000 = 0.0000034 = 0.00034%. That means that the area under the normal curve within $\pm 6\ \sigma$ is 99.99966% with a defect area of 0.00034%.

The following tools are the most common Six Sigma tools, and the rest of the chapter will explain how they are to be used in the concept of sustainability.

Project Charter
SIPOC
Kano Model
CTQ
Affinity Diagram
Measurement Systems Analysis
Gauge R&R
Variation
Graphical Analysis
Location and Spread
Process Capabilities
Cause-and-Effect Diagram
FMEA
Process Mapping
Hypothesis Testing
ANOVA
Correlation
Linear Regression
Theory of Constraints
SMED (Single-Minute Exchange of Dies)
Total Productive Maintenance (TPM)
Design for Six Sigma
Quality Function Deployment
DOE (Design of Experiments)
Control Charts
Control Plan

Project Charter

A project charter is a definition of the project that includes the following:

- Provides problem statement
- Overview of scope, participants, goals, and requirements
- Provides authorization of a new project
- Identifies roles and responsibilities

Once the project charter is approved, it should not be changed.

A project charter begins with the project name, the department of focus, the focus area, and the product or process.

An example of the table of contents for a project charter is shown in Figure 5.2.

A project charter serves as the focus point throughout the project to ensure that the project is on track and the proper people are participating and being held accountable.

The importance of a project charter with respect to sustainability is the living document to educate and give governance for a new project. Sustainability needs to utilize a great deal of education while giving goals and objectives. A project charter will serve as this living document for organizations with specified approaches.

SIPOC

The SIPOC identifies the following:

1. Major tasks and activities
2. The boundaries of the process
3. The process outputs
4. Who receives the outputs (the customers)
5. What the customer requires of the outputs
6. The process inputs
7. Who supplies the inputs (suppliers)
8. What the process requires of the inputs
9. The best metrics to measure

TABLE OF CONTENTS

FIGURE 5.2
Project charter example.

SIPOC stands for the following, shown in Figure 5.3:

Supplier—Know and work with your supplier while making your supplier improve.

Input—Strive to continually improve the inputs by trying to do the right thing the first time.

Process—Describe the process at a high level, but with enough detail to demonstrate to an executive or manager. Understand the process fully by knowing it 100%. Eliminate any mistakes by doing a Poka-Yoke.

FIGURE 5.3
SIPOC.

Output—Strive to continually improve the outputs by utilizing metrics.

Customer—Keep the customer's requirements in sight by remembering they are the most important aspect of the project. The customer makes the specifications; keep the CTQs of the customer in mind.

The following are the SIPOC steps:

1. Gain top-level view of the process.
2. Identify the process in simple terms.
3. Identify external inputs such as raw materials, employees, etc.
4. Identify the customer requirements, also known as outputs.
5. Make sure to include all value-added and non-value-added steps.
6. Include both process and product output variables.

SIPOC implies that the process is understood and helps easily identify opportunities for improvement.

A SIPOC is important in concepts of sustainability because it helps develops a solution for development. Normally the process is mapped out in a well-defined but high level first.

The following is an example of a SIPOC for the steps to recycling a product:

1. Collect and process goods into a container that is friendly to deposit/refund or pick up programs.
2. Undergo a recycling loop where products are sorted by type and parts.
3. Parts go to a smelting plant where parts are put back to original state.
4. Melted parts are sent back to a factory to make new products.
5. Pieces are put back together in a completed product form at a factory.
6. New products are sold.
7. Supplier provides new product.

The SIPOC map is continued with the outputs:

Product ready for recycling
Sorted product
Melted product

Recycled product ready for production
New product

The same process occurs with the inputs:

Newspapers
Plastic
Cardboard
Glass
Magazines
Cans
Metal

The suppliers are then identified:

ABC local newspaper
Coca Cola bottles
Box for ABC macaroni container
Glass jar for ABC sauce
ABC magazine
ABC canned soup

Finally, the customer is realized. It is important to understand the customer is not always the end customer and is also part of the recycling process.

Trash recycling pickup
Container for recycling
Recycling factory
Sorting personnel
Melting personnel
Completed smelting process organizer
Delivery of recycled product to factory
Factory of re-production
Customer buying product

The important part of a SIPOC is to look at the details of the current state and see what improvements can be made for future states. Adding specifications for any of the inputs can identify gaps in the process. Benchmarking one process to another will also identify gaps.

Kano Model

The Kano model was developed by Noriaki Kano in the 1980s. The Kano model is a graphical tool that further categorizes VOC and CTQs into three distinct groups:

- Must-haves
- Performance
- Delighters

The Kano helps identify CTQs that add incremental value versus those that are simply requirements where having more is not necessarily better.

The Kano model (Figure 5.4) engages customers by understanding the product attributes that are most likely important to customers. The purpose of the tool is to support product specifications that are made by the customer and promote discussion while engaging team members. The model differentiates features of products rather than customer needs by understanding necessities and items that are not required whatsoever. Kano also produced a methodology for mapping consumer responses with questionnaires that focused on attractive qualities through reverse qualities. The five categories for customer preferences are as follows:

- Attractive qualities are those that provide satisfaction when fulfilled; however, they do not result in dissatisfaction if not fulfilled.
- One-dimensional qualities are those that provide satisfaction when fulfilled, and dissatisfaction if not fulfilled.

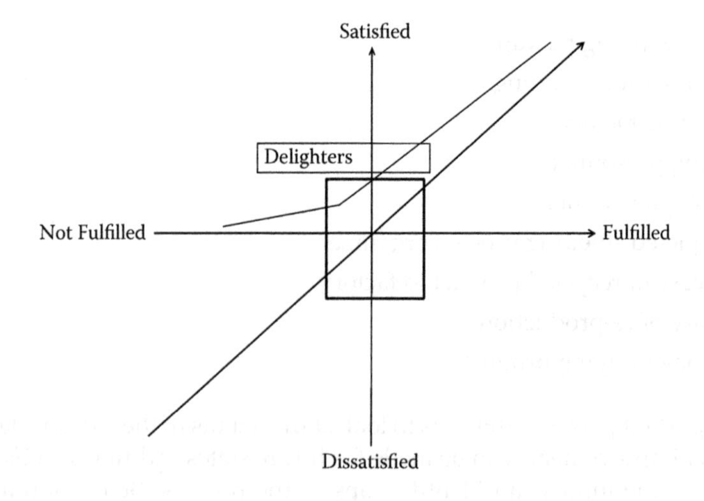

FIGURE 5.4
Kano model.

TABLE 5.2
Kano Table

Answers to a positively formulated question	Answers to a negatively formulated question			
	I like that	That's normal	I don't care	I don't like that
I like that		Delighter	Delighter	Satisfier
That's normal				Dissatisfier
I don't care				Dissatisfier
I don't like that				

- Must-be qualities are those that are taken for granted if fulfilled, but provide dissatisfaction when not fulfilled.
- Indifferent qualities are those that are neither good nor bad resulting in neither customer satisfaction nor dissatisfaction.
- Reverse qualities are those that result in high levels of dissatisfaction from some customers and show that most customers are not alike.

A Kano table is shown in Table 5.2. The Kano model is important to use when being sustainable because it is important to differentiate which aspects we must accomplish to protect our environment and which aspects we can gradually improve upon.

CTQ

CTQ or Critical to Quality are characteristics that are important to the customer. They come from the Voice of the Customer (VOC). CTQs are measureable and quantifiable metrics that come from the VOC. An affinity diagram is an organizational tool for VOCs.

CTQ is critical to sustainability because we need to understand the critical aspects to the environment that matter most. Referring back to Chapter 1, the following are the main areas of consumption:

- Electricity—#1 source of resource consumption
- Natural gas—#2 source of resource consumption
- Water/sewer—#3 source of resource consumption

Therefore, the CTQ characteristics are electricity, natural gas, and water/sewer.

Utilizing a VOC for manufacturing internally is a good way to understand processes that the employees know a great deal about. Therefore, the

TABLE 5.3

CTQ Questions for Internal Customers

	Customer Answer
Why is there no meat in the tray?	
(a) Machinery Problems	10
(b) Product is too tempered out	5
(c) Product is not meant for the machinery	2
Why is there not enough meat in the tray?	
(a) The weights are not dialed in correctly	12
(b) Product is not meant for the machinery	4
(c) There are too many small pieces of product	1

production worker(s) are the customers and the questions are given to them. The questions in Table 5.3 are examples of questions to ask the customers.

The questions can then be put into a pie diagram (see Figure 5.5) to understand what is critical to quality to the customer.

Based on the two questions asked and the responses, the machinery is the CTQ attribute since it affects more than one of the problems.

Affinity Diagram

An affinity diagram is a tool conducted to place large amounts of information into an organized manner by grouping the data into characteristics. The steps for an affinity diagram are as follows:

- Step 1: Clearly define the question or focus of the exercise.
- Step 2: Record all participant responses on note cards or Post-it notes.
- Step 3: Lay out all note cards or post the Post-it's onto a wall.
- Step 4: Look for and identify general themes.
- Step 5: Begin moving the note cards or Post-it notes into the themes until all responses are allocated.
- Step 6: Reevaluate and make adjustments.

An example of an affinity diagram for a basic process is shown in Table 5.4.

The exact same methodology for a basic process can be done for a manufacturing or business process where sustainability is in question. The table is shown in Table 5.5.

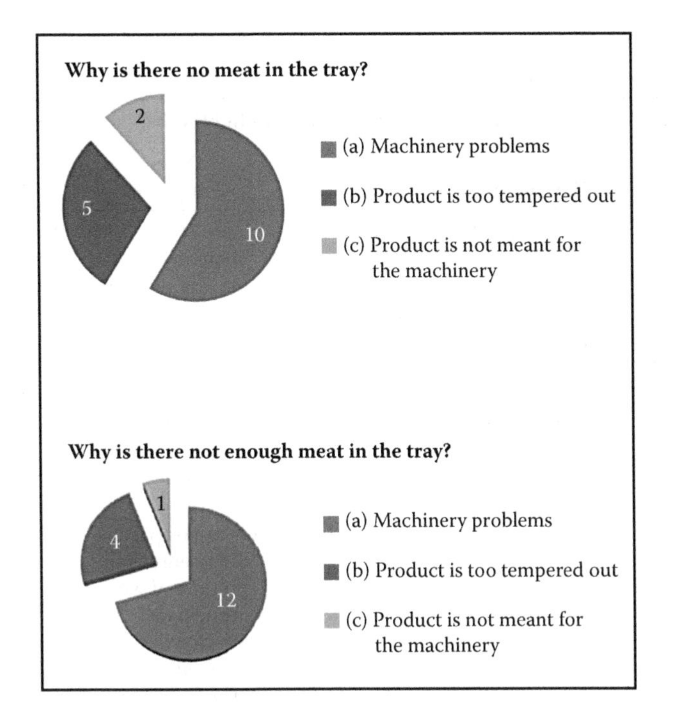

FIGURE 5.5
CTQ based on VOC.

TABLE 5.4
Affinity Diagram

Why do I like my job?				
Management/Personnel	**Standard Processes**	**Pay**	**Company**	**Resources**
My boss gives me freedom to make my own schedule	I get to make my own training tools because we have no standard processes	I get paid well	Family feel	Budget of X dollars
I have made some great friends/met some great coworkers	I get to start from scratch	I get decent amount of vacation		Engineers working under group
I am directing the department and people consider me very intelligent	I like to and get to train others	I get to work from home most Fridays		

TABLE 5.5

Sustainability Affinity Diagram

Should we use energy-efficient lightbulbs?				
Cost	**Environment**	**Looks**	**Other Companies**	**Resources**
It will save money by using less energy	It is good for the environment	The lights look better in the factories	Every other company is already doing it	It is in our budget to make maintenance repairs
The cost for replacing all the lightbulbs will be more expensive than the savings	The lightbulbs last longer than regular light bulbs which will help us from changing them out as often	The lights are not bright enough	We will not sustain continuous improvement without making improvements	The maintenance crew can replace all the lightbulbs
The return on investment will have a one-year turnaround			We can be part of the Energy Star program if we participate, which will gain recognition for ourselves	

The pros and cons are then sought after to have a decision. The decision should be made by having a consensus from the group where the pros outweigh the cons.

Measurement Systems Analysis

Gauge R&R

Gauge R&R is a measurement systems analysis (MSA) technique that uses continuous data based on the following principles:

- Data must be in statistical control.
- Variability must be small compared to product specifications.
- Discrimination should be about one-tenth of product specifications or process variations.
- Possible sources of process variation are revealed by measurement systems.
- Repeatability and reproducibility are primary contributors to measurement errors.

- The total variation is equal to the real product variation plus the variation due to the measurement system.
- The measurement system variation is equal to the variation due to repeatability plus the variation due to reproducibility.
- Total (observed) variability is an additive of product (actual) variability and measurement variability.

Discrimination is the number of decimal places that can be measured by the system. Increments of measure should be about one-tenth of the width of a product specification or process variation that provides distinct categories.

Accuracy is the average quality near to the true value.

The *true value* is the theoretically correct value.

Bias is the distance between the average value of the measurement and the true value, the amount by which the measurement instrument is consistently off target, or systematic error. *Instrument accuracy* is the difference between the observed average value of measurements and the true value. Bias can be measured based by instruments or operators. Operator bias occurs when different operators calculate different detectable averages for the same measure. Instrument bias results when different instruments calculate different detectable averages for the same measure.

Precision encompasses total variation in a measurement system, the measure of natural variation of repeated measurements, and repeatability and reproducibility.

Repeatability is the inherent variability of a measurement device. It occurs when repeated measurements are made of the same variable under absolutely identical condition (same operators, setups, test units, environmental conditions) in the short term. Repeatability is estimated by the pooled standard deviation of the distribution of repeated measurements and is always less than the total variation of the system.

Reproducibility is the variation that results when measurements are made under different conditions. The different conditions may be operators, setups, test units, or environmental conditions in the long term. Reproducibility is estimated by the standard deviation of the average of measurements from different measurement conditions.

The *measurement capability index* is also known as the precision-to-tolerance (P/T) ratio. The equation is ($P/T = 5.15 \times \sigma MS$)/tolerance. The P/T ratio is usually expressed as a percentage and indicates what percentage of the tolerance is taken up by the measurement error. It considers both repeatability and the reproducibility. The ideal ratio is 8% or less; an acceptable is ratio 30% or less. The 5.15 standard deviation accounts for 99% of MS variation and is an industry standard.

The P/T ratio is the most common estimate of measurement system precision. It is useful for determining how well a measurement system can perform with respect to the specifications. The specifications, however, may be inaccurate or need adjustment. The %R&R = (σMS/σTotal) × 100 formula addresses the percentage of the total variation taken up by measurement error and includes both repeatability and reproducibility.

A Gauge R&R can also be performed for discrete data also known as binary data. This data is also known as yes/no or defective/nondefective type data. The data still requires at least 30 data points. The percentages of repeatability, reproducibility, and compliance should be measured. If no repeatability is able to be shown, there will also be no reproducibility. The matches should be above 90% for the evaluations. A good measurement system will have a 100% match for repeatability, reproducibility, and compliance.

If the result is below 90%, the operational definition must be revisited and redefined. Coaching, teaching, mentoring, and standard operating procedures should be reviewed, and the noise should be eliminated.

A Gauge R&R is shown in Figure 5.6 where there is a decision to be made on what equipment is sustainable and what employees are sustainable in a factory that produces butter crèmes.

The Gauge R&R bars are desired to be as small as possible, driving the Part-to-Part bars to be larger.

The averages of each operator is different, meaning the reproducibility is suspect. The operator is having a problem making consistent measurements.

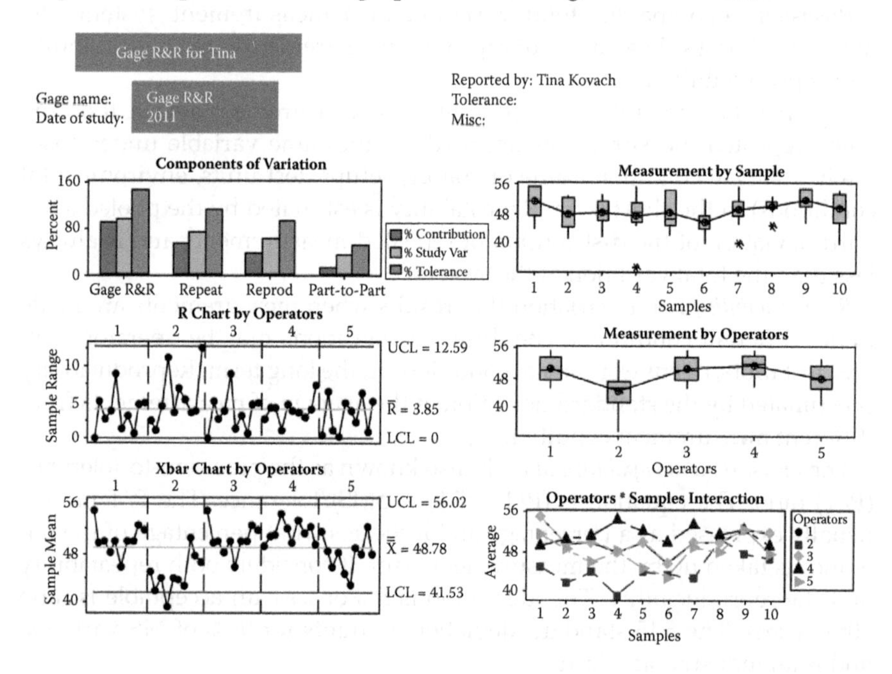

FIGURE 5.6
Gauge R&R.

The Operator*Samples interactions lines should be reasonably parallel to each other. The operators are not consistent to each other.

The Measurement by Samples graph shows there is minimal spread for each sample, and a small amount of shifting between samples.

The Measurement by Operators (Table 5.6) shows that the operators are not consistent, and Operator 2 is normally lower than the rest.

The Sample times Operator of 0.706 shows that the interaction was not significant, which is what is wanted from this study.

The % contribution part to part of 10.81 shows the parts are the same.

The Total Gauge R&R% Study Variation of 94.44, % Contribution of 89.19, Tolerance of 143.25, and Distinct Categories of 1 showed that there was no repeatability or reproducibility and that it was not a good gauge. The number of categories being less than 2 shows that the measurement system is of minimal value since it will be difficult to distinguish one part from another.

The gauge is a bad representation based on Figure 5.7 and Figure 5.8.

The Gauge Run Chart shows that there is no consistency between measurements.

Conclusion: B1 is the best Blender, and Dominic is the best operator. There is no reproducibility or repeatability between any of the measurements.

Variation

Variation is present in all processes, but the goal is to reduce the variation while understanding the root cause of where the variation comes from in the process and why. For Six Sigma to be successful the processes must be in control statistically and the processes must be improved by reducing the variation. The distribution of the measurements should be analyzed to find the variation and depict the outliers or patterns.

The study of variation began with Dr. W. E. Deming, who was also known as the Father of Statistics. Deming stated that variation happens naturally, but the purpose is to utilize statistics to show patterns and types of variations. There are two types of variations that are sought after, special cause variation and common cause variation. Special cause variation refers to out-of-the-ordinary events such as a power outage, whereas common cause variation is inherent in all processes and is typical. The variation is sought to be reduced so that the processes are predictable, in statistical control, and have a known process capability. A root cause analysis should be done on special cause variation so that the occurrence is not to happen again. Management is in charge of common cause variation where action plans are given to reduce the variation.

Assessing the location and spread are important factors as well. Location is known as the process being centered along with the process requirements.

TABLE 5.6

Gage R&R Results

Gage R&R Study—ANOVA Method

Gage R&R for Measurement

Gage name: White Butter Creme Gage R and R
Date of study: 11/18/10
Reported by: Tina Kovach
Tolerance:
Misc:

Two-Way ANOVA Table with Interaction

Source	DF	SS	MS	F	P
Samples	9	282.49	31.388	3.3908	0.004
Operators	4	611.14	152.785	16.5050	0.000
Samples * Operators	36	333.25	9.257	0.8398	0.706
Repeatability	50	551.13	11.023		
Total	99	1778.01			

Alpha to remove interaction term = 0.25

Two-Way ANOVA Table without Interaction

Source	DF	SS	MS	F	P
Samples	9	282.49	31.388	3.0523	0.003
Operators	4	611.14	152.785	14.8573	0.000
Repeatability	86	884.38	10.283		
Total	99	1778.01			

Gage R&R

Source	VarComp	%Contribution (of VarComp)
Total Gage R&R	17.4086	89.19
Repeatability	10.2835	52.68
Reproducibility	7.1251	36.50
Operators	7.1251	36.50
Part-to-Part	2.1104	10.81
Total Variation	19.5190	100.00

Process tolerance = 15

Study		Var %Study		Var %Tolerance
Source	StdDev (SD)	(5.15 * SD)	(%SV)	(SV/Toler)
Total Gage R&R	4.17236	21.4876	94.44	143.25
Repeatability	3.20679	16.5150	72.58	110.10
Reproducibility	2.66929	13.7468	60.42	91.65
Operators	2.66929	13.7468	60.42	91.65
Part-to-Part	1.45274	7.4816	32.88	49.88
Total Variation	4.41803	22.7529	100.00	151.69

Number of Distinct Categories = 1

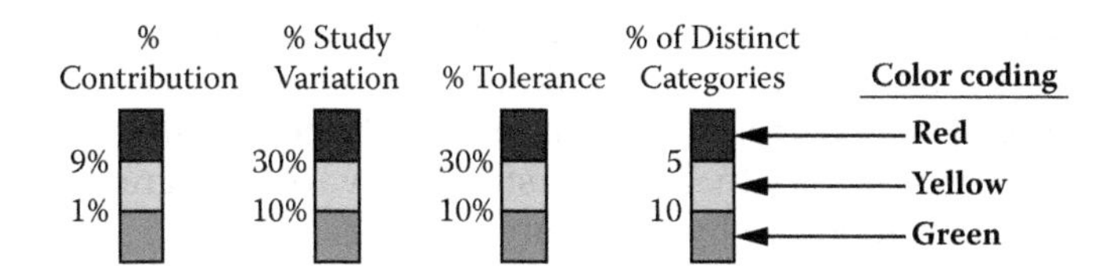

FIGURE 5.7
Gauge R&R results.

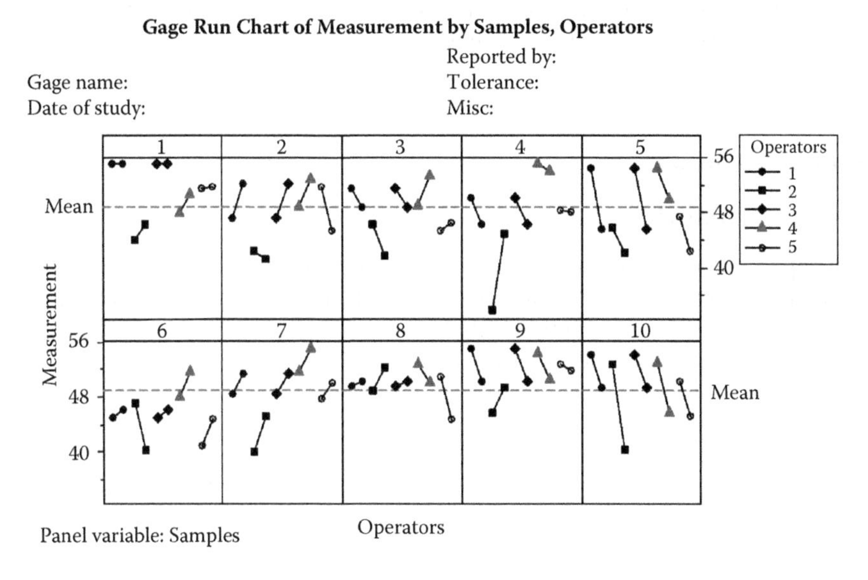

FIGURE 5.8
Gauge run chart.

Spread is known as the observed values compared to the specifications. The stability of the process is required. The process is said to be in statistical control if the distribution of the measurements have the same shape, location, and spread over time. This is the point in time where all special causes of variation are removed and only common cause variation is present.

An *average, central tendency* of a data set is a measure of the "middle" or "expected" value of the data set. Many different descriptive statistics can be chosen as measurements of the central tendency of the data items. These include the arithmetic mean, the median, and the mode. Other statistical measures such as the standard deviation and the range are called measures of spread of data. An average is a single value meant to represent a list of values. The most common measure is the arithmetic mean but there are many other measures of central tendency such as the median (used most often when the distribution of the values is skewed by small numbers with very high values).

As stated before, special cause variation would be occurrences such as power outages, large mechanical breakdowns, and so on. Common cause variations would be occurrences such as electricity being different by a few thousand kilowatts per month. To understand the variation, graphical analyses should be done, followed by capability analyses.

It is important to understand the variation in the systems so that the best-performing equipment is used. The variation sought after is in turn utilized for sustainability studies. The best-performing equipment should be utilized the most and the least-performing equipment should be brought back to its original state of condition and then upgraded or fixed to be capable. Capability indices are explained next.

Process Capabilities

The capability of a process is the spread that contains most of the values of the process distribution. Capability can only be established on a process that is stable with a distribution that only has common cause variation. Figure 5.9 shows an example of process capability.

Capable Process (C_p)

A process is capable ($C_p \geq 1$) if its natural tolerance lies within the engineering tolerance or specifications. The measure of process capability of a stable

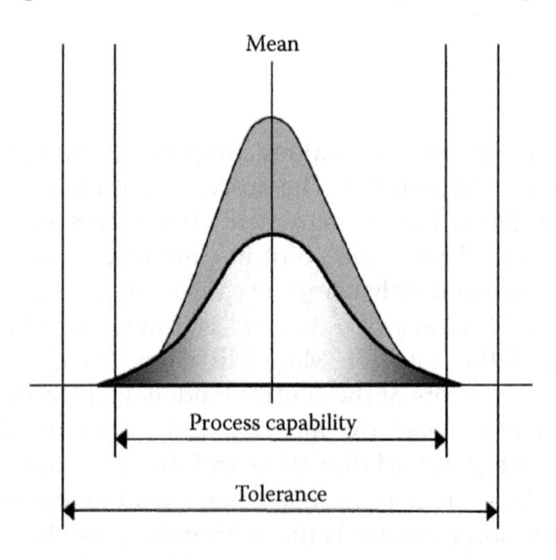

FIGURE 5.9
Process capability example.

process is $6\hat{\sigma}$, where $\hat{\sigma}$ is the inherent process variability that is estimated from the process. A minimum value of $C_p = 1.33$ is generally used for an ongoing process. This ensures a very low reject rate of 0.007% and therefore is an effective strategy for prevention of nonconforming items. C_p is defined mathematically as

$$C_p = \frac{USL - LSL}{6\hat{\sigma}}$$

$$= \frac{\text{allowable process spread}}{\text{actual process spread}}$$

where

USL = upper specification limit

LSL = lower specification limit

C_p measures the effect of the inherent variability only. The analyst should use R-bar/d_2 to estimate $\hat{\sigma}$ from an R-chart that is in a state of statistical control, where R-bar is the average of the subgroup ranges and d_2 is a normalizing factor that is tabulated for different subgroup sizes (n). We don't have to verify control before performing a capability study. We can perform the study and then verify control after the study with the use of control charts. If the process is in control during the study, then our estimates of capabilities are correct and valid. However, if the process was not in control, we would have gained useful information, as well as proper insights as to the corrective actions to pursue.

Capability Index (C_{pk})

Process centering can be assessed when a two-sided specification is available. If the capability index (C_{pk}) is equal to or greater than 1.33, then the process may be adequately centered. C_{pk} can also be employed when there is only one-sided specification. For a two-sided specification, it can be mathematically defined as

$$C_{pk} = \text{Minimum}\left\{\frac{USL - \bar{X}}{3\hat{\sigma}}, \frac{\bar{X} - LSL}{3\hat{\sigma}}\right\}$$

where $\bar{X}$ = overall process average

However, for a one-sided specification, the actual C_{pk} obtained is reported. This can be used to determine the percentage of observations out of specification. The overall long-term objective is to make C_p and C_{pk} as large as possible by continuously improving or reducing process variability, $\hat{\sigma}$, for every iteration so that a greater percentage of the product is near the key quality characteristics target value. The ideal is to center the process with zero variability.

If a process is centered but not capable, one or several courses of action may be necessary. One of the actions may be that of integrating designed experiment to gain additional knowledge on the process and in designing control strategies. If excessive variability is demonstrated, one may conduct a nested design with the objective of estimating the various sources of variability. These sources of variability can then be evaluated to determine what strategies to use to reduce or permanently eliminate them. Another action may be that of changing the specifications or continuing production and then sorting the items. Three characteristics of a process can be observed with respect to capability, as summarized below:

1. The process may be centered and capable.
2. The process may be capable but not centered.
3. The process may be centered but not capable

Possible Applications of the Process Capability Index

The potential applications of process capability index are summarized below:

- *Communication:* C_p and C_{pk} have been used in industry to establish a dimensionless common language useful for assessing the performance of production processes. Engineering, quality, manufacturing, and so on, can communicate and understand processes with high capabilities.
- *Continuous improvement:* The indices can be used to monitor continuous improvement by observing the changes in the distribution of process capabilities. For example, if there were 20% of processes with capabilities between 1 and 1.67 in a month, and some of these improved to between 1.33 and 2.0 the next month, then this is an indication that improvement has occurred.
- *Audits:* There are so many various kinds of audits in use today to assess the performance of quality systems. A comparison of in-process capabilities with capabilities determined from audits can help establish problem areas.
- *Prioritization of improvement:* A complete printout of all processes with unacceptable C_p or C_{pk} values can be extremely powerful in establishing the priority for process improvements.
- *Prevention of nonconforming product:* For process qualification, it is reasonable to establish a benchmark capability of $C_{pk} = 1.33$, which will make nonconforming products unlikely in most cases.

Potential Abuse of C_p and C_{pk}

In spite of its several possible applications, the process capability index has some potential sources of abuse as summarized below:

- Problems and drawbacks: C_{pk} can increase without process improvement even though repeated testing reduces test variability; the wider the specifications, the larger the C_p or C_{pk}, but the action does not improve the process.
- Analysts tend to focus on number rather than on process.
- *Process control:* Analysts tend to determine process capability before statistical control has been established. Most people are not aware that capability determination is based on process common cause variation and what can be expected in the future. The presence of special causes of variation makes prediction impossible and capability index unclear.
- *Non-normality:* Some processes result in non-normal distribution for some characteristics. Since capability indices are very sensitive to departures from normality, data transformation may be used to achieve approximate normality.
- *Computation:* Most computer-based tools do not use $\overline{R}/d_2$ to calculate σ.

When analytical and statistical tools are coupled with sound managerial approaches, an organization can benefit from a robust implementation of improvement strategies. One approach that has emerged as a sound managerial principle is "lean," which has been successfully applied to many industrial operations.

C_p and C_{pk} are capability analyses that can only be done with normal data. It is very easy to use any data for capability analyses especially on software systems that will calculate the data automatically. The first step in doing the capability analysis is to have continuous data and check for normality. Only if the data is normal can the capability studies be done. If the data is not normal, the special cause variation is sought after. Data points may only be taken out if the reasoning is known for the data point that is an outlier (e.g., temperature change, shift change). Once an outlier is found for a known reason, the outlier can be removed and the data can be checked for normality once again. If there is no root cause for the outlier, more data must be taken, but capability analyses should not be done until the normality is proven.

The importance of finding the capable equipment or products in a business through process capabilities will allow the variation to be found through benchmarking. The best processes should be used for these benchmarking techniques. The best-in-class (BIC) practices should be performed on the

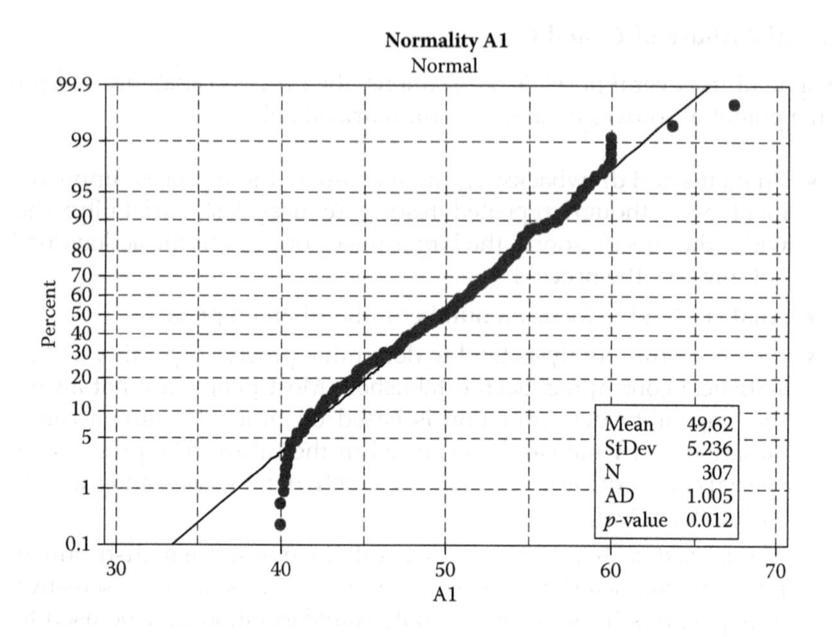

FIGURE 5.10
Normality of equipment A.

different equipments/products/processes. Then improvements should be made on the areas that are not as capable. It is very important to perform preventative maintenance on all and any equipment in order for the equipment to stay performing at the highest possible process capability.

When two or more equipment pieces are being compared, the first step is to perform a normality test as stated above. See Figures 5.10 through 5.15 for normality test examples.

The conclusions that come from the normality tests are the following:

Blender A1 is not normal.

Blender B1 is the most normal.

Blender B2 is not normal.

Small Mixer is JUST in at normal.

The normal pieces of equipment are then checked for the process capabilities:

Blender B1 is your best blender, the short-term capability (Cp of 1.34) is approximately equivalent to a short-term *Z* of 4, which is good. The long-term capability needs some improvement (Ppk of 0.94).

Small Mixer is just normal, but is better than Blender A1 or B2. The short-term capability (Cp of 0.69) needs improvement along with the long-term capability (Ppk of 0.42).

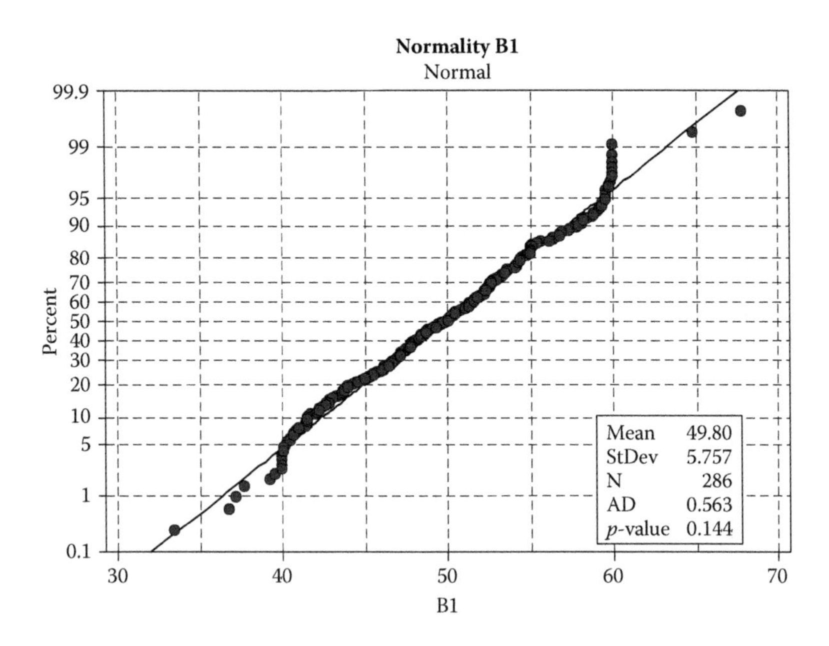

FIGURE 5.11
Normality of equipment B.

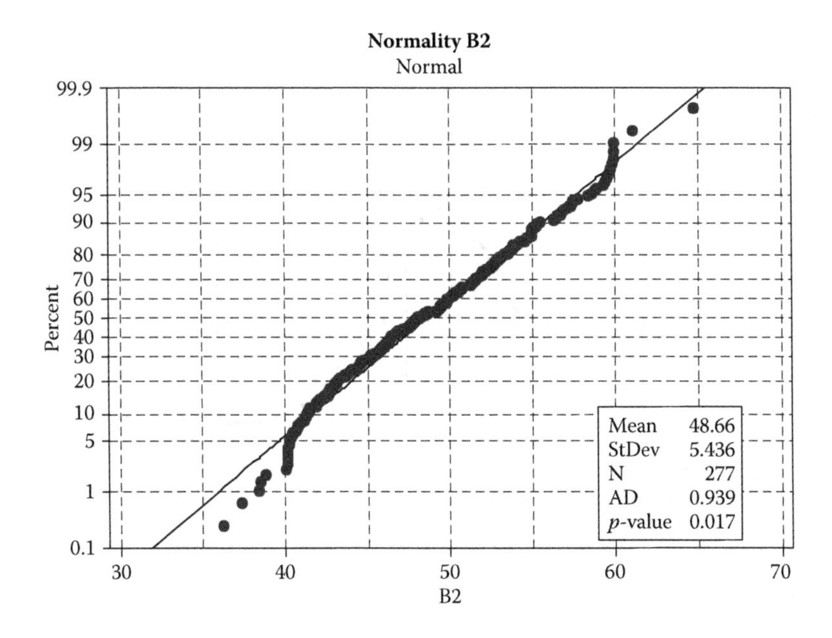

FIGURE 5.12
Normality of equipment B2.

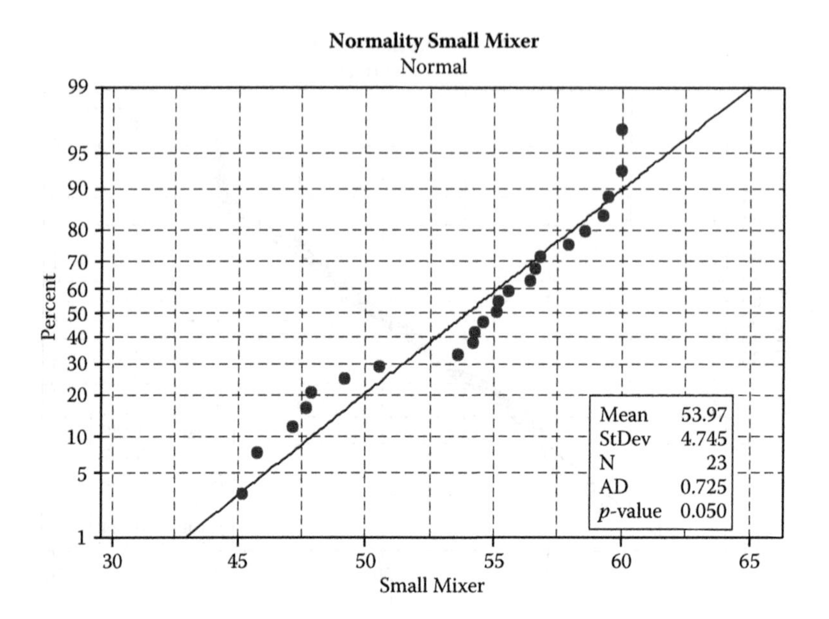

FIGURE 5.13
Normality of equipment small mixer.

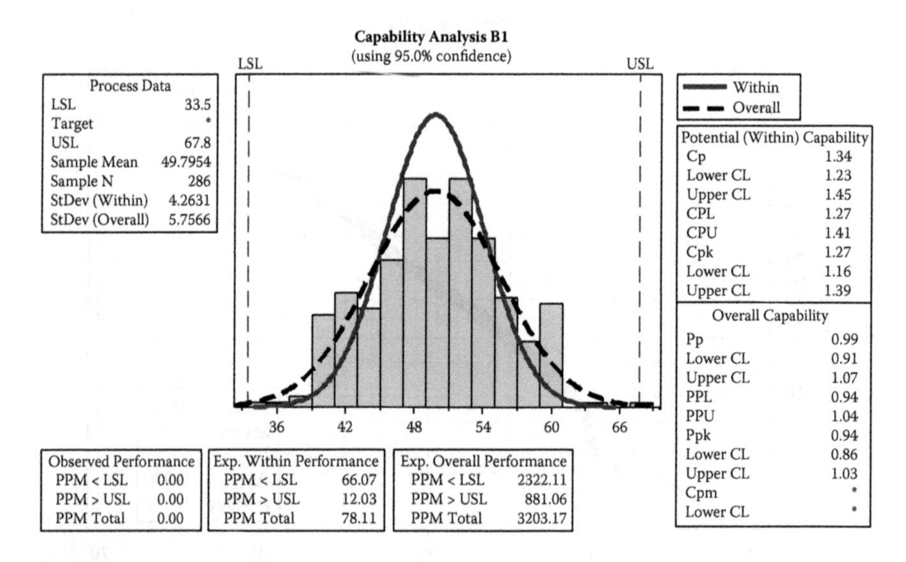

FIGURE 5.14
Process capability B1.

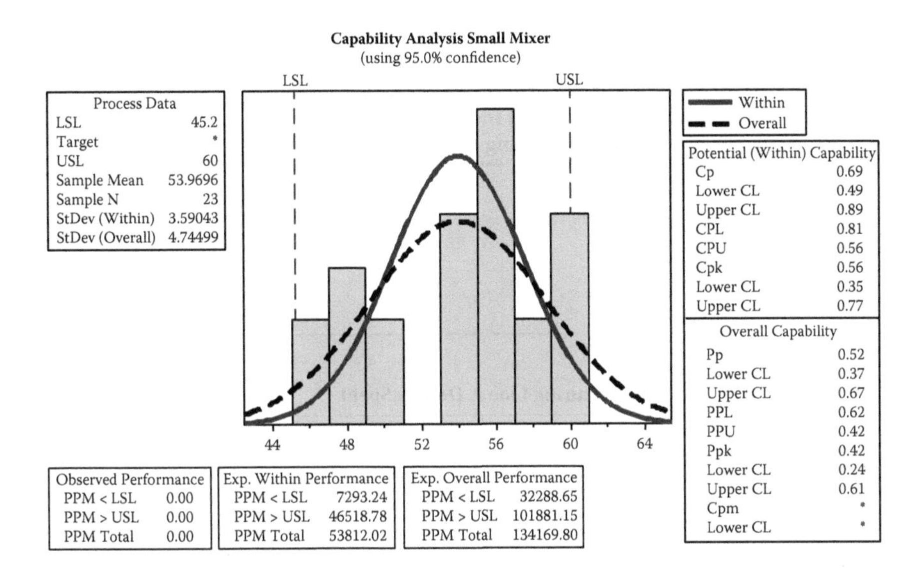

FIGURE 5.15
Process capability small mixer.

The process capability analyses should be continued in a systematic fashion (i.e., monthly or quarterly) to understand if the processes are improving. Continuous improvement should be performed on the equipment for the best capabilities.

Graphical Analysis

Graphical analyses are visual representations of tools that show meaningful key aspects of projects. These tools are commonly known as dotplots, histograms, normality plots, Pareto diagrams, second-level Paretos (also known as stratification), boxplots, scatter plots, and marginal plots. The plotting of data is a key beginning step to any type of data analysis because it is a visual representation of the data.

If a particular manufacturing company wants to understand where the majority of their electrical costs are coming from while trying to reduce those costs, the steps are as follows:

Gather the data for where the costs are coming from, shown in Table 5.7.

Pareto the costs to understand the largest hitters, shown in Figure 5.16.

The largest cost can be clearly seen for a next step in the process.

TABLE 5.7

Manufacturing Electrical Costs

Largest Areas of Cost for Electricity in Manufacturing Plant	Dollar Spent
Freezing	$500,000
Manufacturing line	$400,000
Cooking	$350,000
Packaging	$125,000
Canning	$100,000

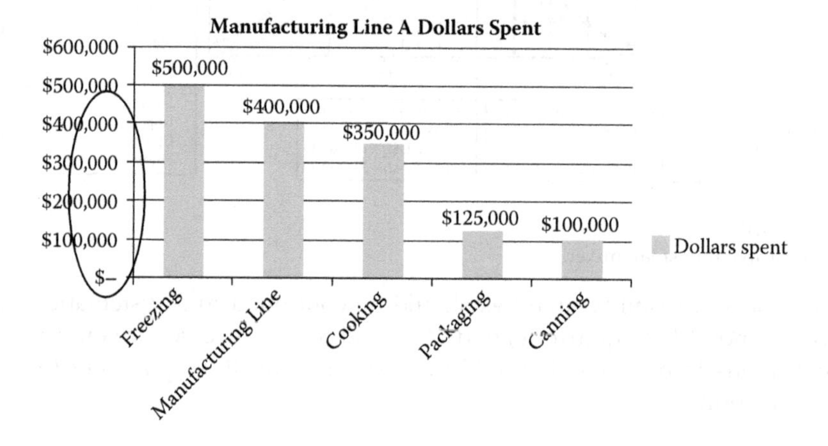

FIGURE 5.16
Pareto of manufacturing costs.

Process Mapping

The importance of process mapping is to depict all functions in the process flow while understanding if the functions are value- or non-value-added. An example is shown in Figure 5.17. Any delays are to be eliminated and decisions are meant to be as efficient as possible. The purpose of process mapping is to have a visual image of the process.

Cause-and-Effect Diagram

After a process is mapped, the cause-and-effect (C&E) diagram can be completed. This process is so important because it completed root cause analysis.

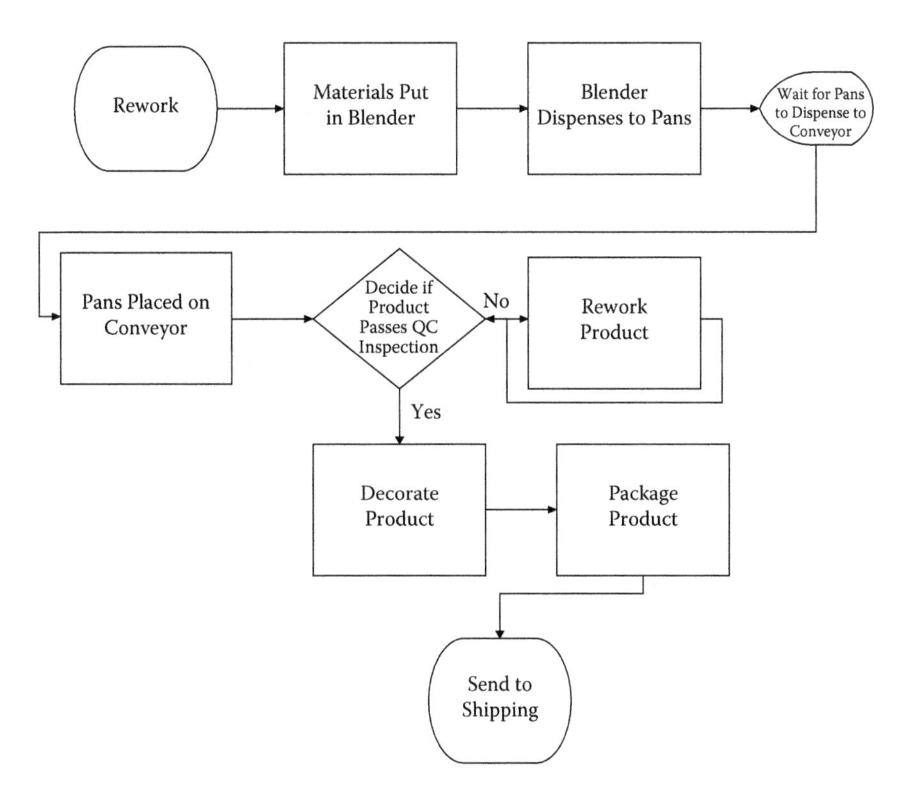

FIGURE 5.17
Process mapping.

The basis behind root cause analysis is to ask, "Why?" five times to get to the actual root cause. Many times problems are "Band-Aided" to fix the top-level problem, but the actual problem itself is not addressed.

A C&E diagram is shown in Figure 5.18. The fishbone is broken out to the most important categories in an environment:

- Measurements
- Material
- Personnel
- Environment
- Methods
- Machines

This process requires a team to do a great deal of brainstorming where they focus on the causes of the problems based on the categories. The "fish head" is the problem statement.

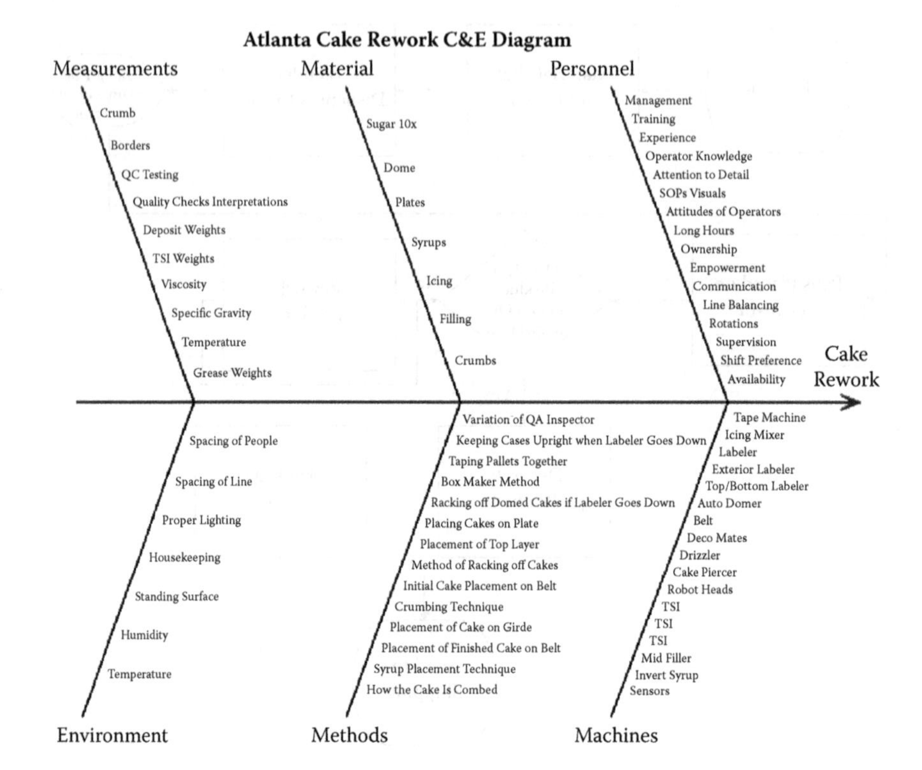

FIGURE 5.18
C&E diagram.

Failure Mode and Effect Analysis (FMEA)

To select action items from the C&E diagram and prioritize the projects, FMEAs are completed. The FMEA will identify the causes, assess risks, and determine further steps. The steps to an FMEA are the following:

1. Define process steps.
2. Define functions.
3. Define potential failure modes.
4. Define potential effects of failure.
5. Define the severity of a failure.
6. Define the potential mechanisms of failure.

7. Define current process controls.
8. Define the occurrence of failure.
9. Define current process control detection mechanisms.
10. Define the ease of detecting a failure.
11. Multiply severity, occurrence, and detection to calculate a risk priority number (RPN).
12. Define recommended actions.
13. Assign actions with key target dates to responsible personnel.
14. Revisit the process after actions have been taken to improve it.
15. Recalculate RPNs with the improvements.

An FMEA is shown in Figure 5.19.

What can be seen from the FMEA that is an important aspect to sustainability is the RPN number reducing after the action items. It is important to understand the process's severity to a customer and increasing the capability of the process to in turn improve the process. The RPN's reducing will make the entire process more sustainable by being able to deliver the process at the best capabilities through thorough project management. It is important to maintain the FMEA so that once a process is improved it is not forgotten about.

Hypothesis Testing

Hypothesis testing validates assumptions made by verification of the processes based on statistical measures. It is important to use at least 30 data points for hypothesis testing so that there is enough data to validate the results. See Table 5.8.

Normality of the data points must be found for the hypothesis testing to be accurate.

The assumptions are shown in the null and alternate hypothesis:

H_0 = (The null hypothesis): The difference is equal to the chosen reference value; $\mu_1 - \mu_2 = 0$

H_a = (The alternate hypothesis): The difference is not equal to the chosen reference value; $\mu_1 - \mu_2$ is not = 0

95% CI for mean difference: (1.16, 6.69) T-Test of mean difference = 0 (vs. not = 0): T-Value = 2.90; p-Value = 0.007

#	Process Function (Step)	Potential Failure Modes (process defects)	Potential Failure Effects (KPOVs)	SEV	Class	Potential Causes of Failure (KPIVs)	OCC	Current Process Controls	DET	RPN	Recommend Actions	Responsible Person & Target Date	Taken Actions	SEV	OCC	DET	RPN
1	Drop batch to votator 1	Water in pipe	Heated Finished Product, Cooling Step 1	8	XX	Piping	9	Open votator valves after flush (may not happen all the time)	8	576	(1) Training to be completed week of 8/16/10 and sign off sheet to be signed off after flushing. (2) Audits completed 1× per week.	(1) Brenda Ellis and Ben Jones, August 20, 2010 (2) Ben K, Ben Jones, and Todd Waiz, August 27, 2010	Training Manual Completed, Audits continued	5	5	5	125
2	Put into Pre-mix kettle	Supplier providing product out of spec	Melted Shortening	8	XX	P112 Shortening	7	No process check	10	560	Check COA every load, every lot	Justin Hollander, Sept 1, 2010	COA's now always checked and supplier notified if issues	1	1	1	1
3	Put into Pre-mix kettle	Quality of shortening	Melted Shortening	8	XX	P112 Shortening	8	No process check	8	512	Check COA every load, every lot	Justin Hollander, Sept 1, 2010	COA's now always checked and supplier notified if issues	1	1	1	1

FIGURE 5.19
FMEA.

TABLE 5.8
Paired t-Test

Paired t for Before–After	N	Mean	St Dev	SE Mean
Before	30	83.623	5.195	0.948
After	30	79.697	4.998	0.913
Difference	30	3.93	7.41	1.35

The confidence interval for the mean difference between the two materials does not include zero, which suggests a difference between them. The small *p*-value ($p = 0.007$) further suggests that the data are inconsistent with H_0: $\mu_1 - \mu_2 = 0$, that is, the two materials do not perform equally. Specifically, the first set (mean = 79.697) performed better than the next set (mean = 83.623) in terms of weight control over the time span. Conclusion: Reject H_0. The difference is not equal to the chosen reference value: $\mu_1 - \mu_2$ is not = 0. The Histogram of Differences and the Boxplot of Differences are graphed in Figure 5.20 and Figure 5.21, respectively.

The confidence interval for the mean difference between the two materials does not include zero, which suggests a difference between them. The small *p*-value ($p = 0.007$) further suggests that the data are inconsistent with H_0: $\mu\, d = 0$; that is, the two materials do not perform equally. Specifically, the first set (mean = 79.697) performed better than the other set (mean = 83.623) in terms of weight control over the time span. Conclusion: Reject H_0. The difference is not equal to the chosen reference value: $\mu_1 - \mu_2$ is not = 0.

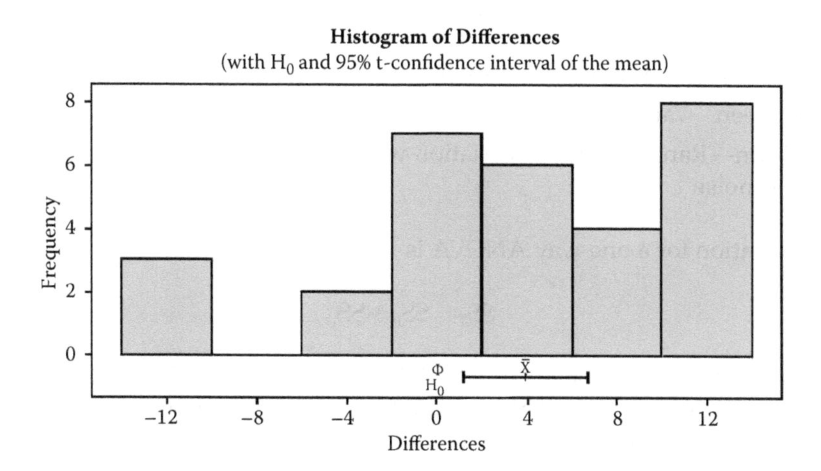

FIGURE 5.20
Histogram of differences.

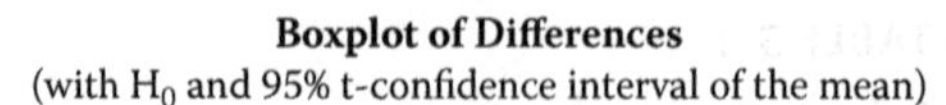

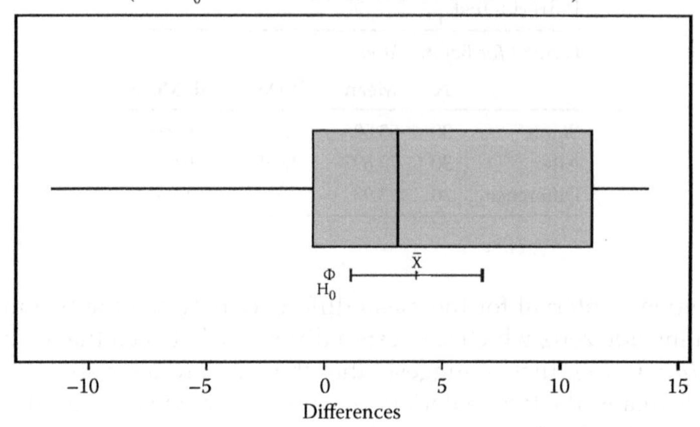

FIGURE 5.21
Boxplot of differences.

ANOVA

The purpose of an ANOVA, also known as analysis of variance, is to determine if there is a relationship between a discrete, independent variable and a continuous, dependent output. There is a one-way ANOVA, which includes one-factorial variance, and a two-way ANOVA, which includes a two-factorial variance. Three sources of variability are sought after:

Total—Total variability within all observations

Between—Variation between subgroup means

Within—Random chance variation within each subgroup, also known as noise

The equation for a one-way ANOVA is

$$SS_T = SS_F + SS_e$$

The principles for the one-way ANOVA and two-way ANOVA are the same except that in a two-way ANOVA, the factors can take on many levels. The total variability equation for a two-way ANOVA is

$$SS_T = SS_A + SS_B + SS_{AB} + SS_e$$

where

SS_T = total sum of squares
SS_F = sum of squares of the factor
SS_e = sum of squares from error
SS_A = sum of squares for factor A
SS_B = sum of squares for factor B
SS_{AB} = sum of squares due to interaction of factors A and B

If the ANOVA shows that at least one of the means is different, a pairwise comparison is done to show which means are different. The residuals, variance, and normality should be examined and the main effects plot and interaction plots should be generated.

The F-ratio in an ANOVA compares the denominator to the numerator to see the amount of variation that is expected. When the F-ratio is small, which is normally close to 1, the value of the numerator is close to the value of the denominator, and the null hypothesis cannot be rejected stating the numerator and denominator are the same. A large F-ratio indicates the numerator and denominator are different, also known as the MS error where the null hypothesis is rejected.

Outliers should also be sought after in the ANOVA showing that the variability is affected.

The main effects plot shows the mean values for the individual factors being compared. The differences between the factor levels can be seen with the slopes in the lines. The *p*-values can help determine if the differences are significant.

Interaction plots show the mean for different combinations of factors.

Correlation

The linear relationship between two continuous variables can be measured through correlation coefficients. The correlation coefficients are values between –1 and 1.

If the value is around 0, there is no linear relationship.
If the value is less than 0.05, there is a weak correlation.
If the value is less than 0.08, there is a moderate correlation.
If the value is greater than 0.08, there is a strong correlation.
If the value is around 1, there is a perfect correlation.

Simple Linear Regression

The regression analysis describes the relationship between a dependent and independent variable as a function $y = f(x)$.

The equation for simple linear regression as a model is

$$Y = b_0 + b_1 x + {}_E$$

where

Y is the dependent variable
b_0 is the axis intercept
b_1 is the gradient of the regression line
x is the independent variable
E is the error term or residuals

The predicted regression function is tested with the following formula:

$$R^2 = \frac{SSTO - SSE}{SSTO}$$

where

$$SSTO = \begin{cases} \underset{\sim}{Y}' - n\overline{Y}^2 & \text{if constant} \\ \underset{\sim}{Y}\underset{\sim}{Y} & \text{if no constant} \end{cases}$$

NOTE: When the no constant option is selected, the total sum of square is uncorrected for the mean. Thus, the R^2 value is of little use, since the sum of the residuals is not zero.

The F-test shows if the predicted model is valid for the population and not just the sample. The model is statistically significant if the predicted model is valid for the population.

The regression coefficients are tested for significance through t-tests with the following hypothesis:

H_0: $b_0 = 0$, the line intersects the origin
H_A: $b_0 \neq 0$, the line does not intersect the origin

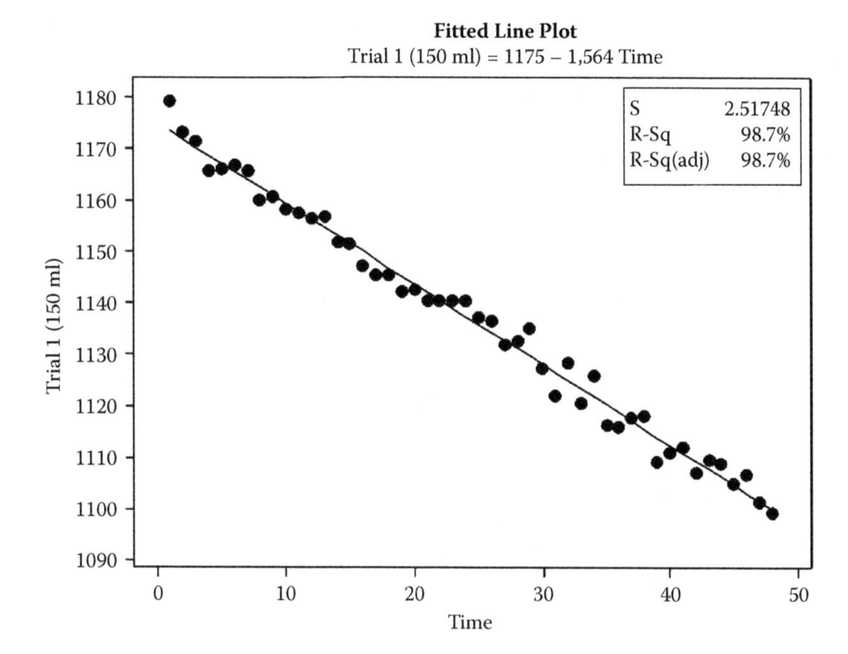

FIGURE 5.22
Fitted line plot.

H_0: $b_1 = 0$, there is no relationship between the independent variable x_i and the dependent variable y

H_A: $b_1 \neq 0$, there is a relationship between the independent variable x_i and the dependent variable y

A fitted line plot can be done to see the inverse relationship as shown in Figure 5.22.

After the inverse relationship is seen, a regression analysis can be performed.

An example is shown below for the analysis of whether there was a pressure degradation over time on a particular piece of equipment:

A linear relationship was sought after. First, it was sought to see if there was correlation since it can be seen that there is a linear relationship between the variables. The y was the measurement and the x was the time.

Correlations: Trial 1, Time

Pearson correlation of Trial 1 (150 ml) and Time = –0.994

p-Value = 0.000

This correlation coefficient of $r = -0.994$ shows a high, positive dependence. The p-value being less than 0.05 also shows that the correlation coefficient is significant. The regression equation is

Trial 1 = 1175 – 1.56 Time

Predictor	Coef	SE Coef	T	P
Constant	1174.89	0.74	1591.47	0.000
Time	–1.56360	0.02623	–59.61	0.000

S = 2.51748 R-Sq = 98.7% R-Sq(adj) = 98.7%

Analysis of Variance

Source	DF	SS	MS	F	P
Regression	1	22522	22522	3553.63	0.000
Residual error	46	292	6		
Total	47	22813			

Unusual Observations

Trial 1						
Obs	**Time (150ml)**	**Fit**	**SE Fit**	**Residual**	**St**	**Resid**
1	1.0	1179.00	1173.33	0.72	5.67	2.35R
29	29.0	1135.10	1129.55	0.38	5.55	2.23R

R denotes an observation with a large standardized residual.

For each time, there is 1.5 measurement of degradation according to the equation:

$$Y_1 = \beta_0 + \beta_1 X_1$$

The slope equals 1.564. There is a negative slope.

In this particular situation, it becomes critical after losing more than 10% of the measurement in the specification range. According to the graph, about every 10 trials, there is a degradation of about 15.

Hypothesis Testing

The p-value shows that this is not normal.

H_0 = Accept null hypothesis ($\beta = 0$, no correlation)

H_a = Reject null hypothesis ($\beta \neq 0$, there is correlation)

Therefore, reject null hypothesis. There is correlation.

In the above, the R^2 value is 98.7%, which means that 98.7% of the Y variable's (pressure) can be explained by the model (the regression equation).

The residuals are then evaluated. See Figures 5.23, 5.24, and 5.25.

The normality is also taken of the residuals.

The normality test passes with a value of 0.850. The residuals are in control.

The residuals are contained in a straight band, with no obvious pattern in the graph showing that this model is adequate.

Conclusion

Reject H_0; the slope of the line does not equal 0. There is a linear relationship in the measurement versus time, showing that there is correlation. This model proves to be adequate due to the testing done above.

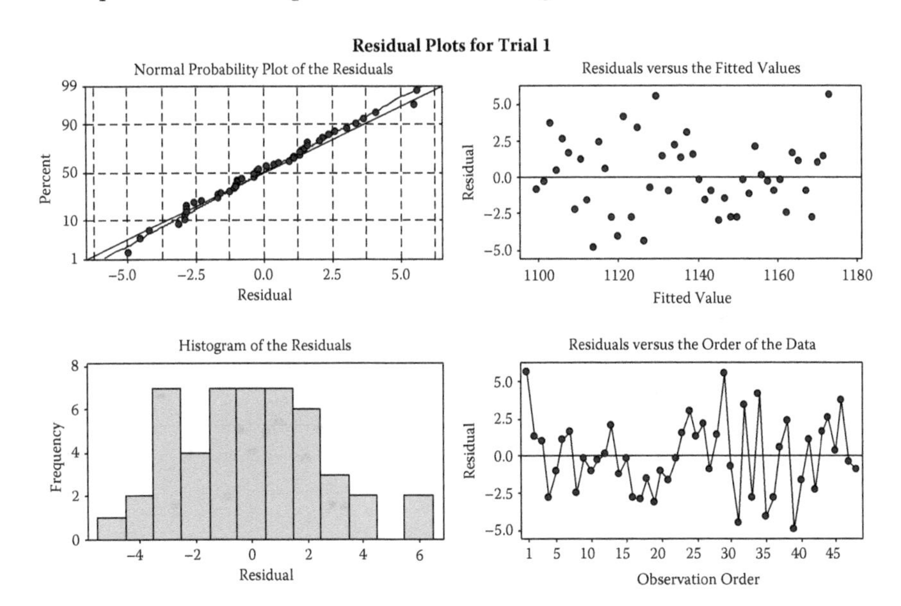

FIGURE 5.23
Residual plots.

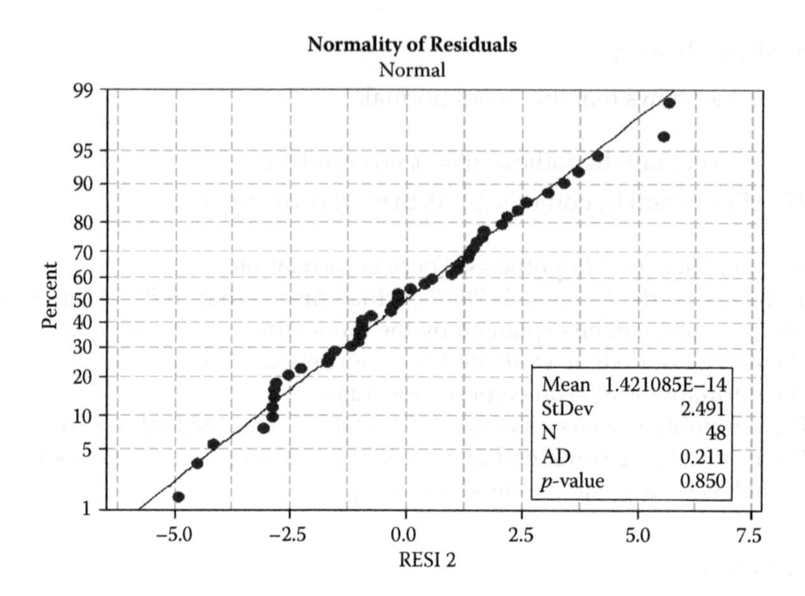

FIGURE 5.24
Normality of residuals.

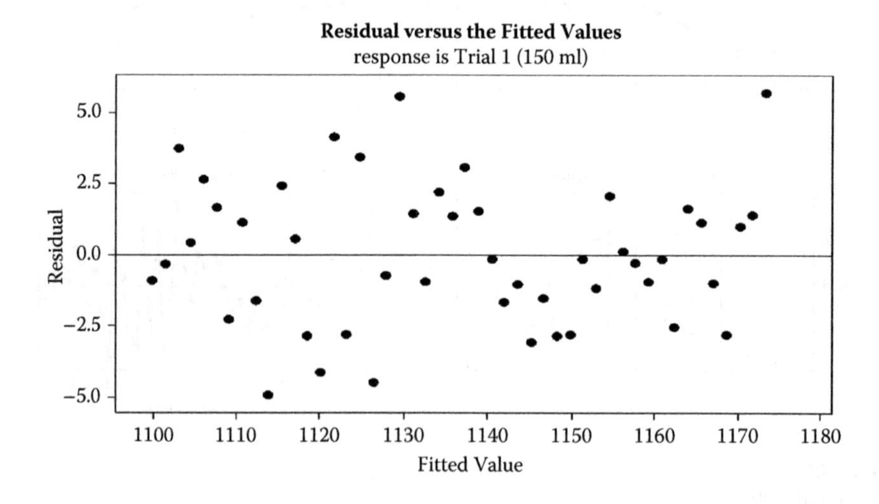

FIGURE 5.25
Residuals vs. fitted values.

Theory of Constraints

Dr. Eliyahu M. Goldratt created a theory of constraints (TOC). This management theory proved that every system has at least one constraint limiting it from 100% efficiency. The analysis of a system will show the boundaries of the system. TOC not only shows the cause of the constraints but also provides a way to resolve the constraints. There are two underlying concepts with TOC:

1. System as chains
2. Throughput, inventory management, and operating expenses

The performance of the entire system is called the chain. The performance of the system is based on the weakest link of the chain or the constraint. The remaining links are known as non-constraints. Once the constraint is improved, the system becomes more productive or efficient, but there is always a weakest link or constraint. This process continues until there is 100% efficiency.

If there are three manufacturing lines, they produce the following:

1. 250 units/day
2. 500 units/day
3. 600 units/day

The weakest link is manufacturing line 1 because it produces the least amount of units/day. The weakest link is investigated until it reaches the capacity of the non-constraints. After the improvement has been made, the new weakest link is investigated until the full potential of the manufacturing lines can be fulfilled without exceeding market demand. If the external demand is fewer than the internal capacity, it is known as an external constraint.

Throughput can be defined in the following formula:

$$\frac{\textit{Sales Price} - \textit{Variable Costs}}{\textit{Time}}$$

Profits should be understood when dealing with throughput.

Inventories are known as raw materials, unfinished goods, purchased parts, or any investments made. Inventory should be seen as dollars on shelves. Any inventory is a waste unless utilized in a just-in-time manner.

Operating expenses should include all expenses utilized to produce a good. The less the operating expenses, the better. These costs should include direct labor, utilities, supplies, and depreciation of assets.

Applying the TOC concept helps guide making the weakest link stronger. There are five steps to the process of TOC:

1. Identify the constraint or the weakest link.
2. Exploit the constraint by making it as efficient as possible without spending money on the constraint or considering upgrades.
3. Subordinate everything else to the constraint—Adjust the rest of the system so the constraint operates at its maximum productivity. Evaluate the improvements to ensure that the constraint has been addressed properly and it is no longer the constraint. If it is still the constraint, complete the steps; otherwise skip step 4.
4. Elevate the constraint—This step is required only if steps 2 and 3 were not successful. The organization should take any action on the constraint to eliminate the problem. This is the process where money should be spent on the constraint or upgrades should be investigated.
5. Identify the next constraint and begin the five-step process over. The constraint should be monitored and continuous improvement should be completed.

Single-Minute Exchange of Die (SMED)

Single-Minute Exchange of Die (SMED) consists of the following:

- Theory and set of techniques to make it possible to perform equipment setup and changeover operations in under 10 minutes
- Originally developed to improve die press and machine tool setups, but principles apply to changeovers in all processes
- It may not be possible to reach the "single-minute" range for all setups, but SMED dramatically reduces setup times in almost every case
- Leads to benefits for the company by giving customers variety of products in just the quantities they need
- High quality, good price, speedy delivery, less waste, cost effective

It is important to understand large lot production, which leads to trouble.

The three key topics to consider when understanding large lot production are the following:

- Inventory waste
    - Storing what is not sold costs money
    - Ties up company resources with no value to the product
- Delay
    - Customers have to wait for the company to produce entire lots rather than just what they want

- Declining quality
  - Storing unsold inventory increases chances of product being scrapped or reworked, adding costs

Once this is realized, the benefits of SMED can be understood:

- Flexibility
  - Meet changing customer needs without excess inventory
- Quicker delivery
  - Small-lot production equals less lead time and less customer waiting time
- Better quality
  - Less inventory storage equals fewer storage-related defects
  - Reduction of setup errors and elimination of trial runs for new products
- Higher productivity
  - Reduction in downtime
  - Higher equipment productivity rate

Two types of operations are realized during setup operations, which consist of internal and external operations. Internal setup is a setup that can only be done when the machine is shut down (e.g., a new die can only be attached to a press when the press is stopped).

External setup is a setup that can be done while the machine is still running (e.g., bolts attached to a die can be assembled and sorted while the press is operating).

It is important to convert as much internal work as possible to external work, which is shown in Figure 5.26.

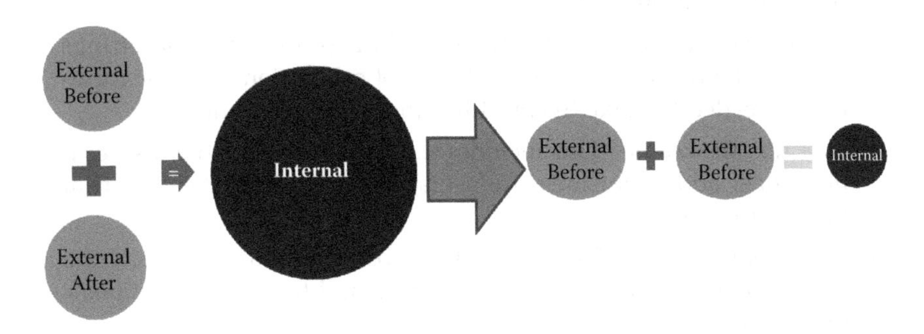

FIGURE 5.26
Internal vs. external setups.

Four important questions to ask yourself when understanding SMED are the following:

- How might SMED benefit your factory?
- Can you see SMED benefiting you?
- What operations are internal operations?
- What operations are external operations?

There are three stages to SMED, which are defined below:

- Separate internal and external setup
    - Distinguish internal vs. external
    - By preparing and transporting while the machine is running can cut changeover times by as much as 50%
- Convert internal setup to external setup
    - Reexamine operations to see whether any steps are wrongly assumed as internal steps
    - Find ways to convert these steps to external setups
- Streamline all aspects of setup operations
    - Analyze steps in detail
    - Use specific principles to shorten time needed especially for steps internally with machine stopped

Five traditional setup steps are also defined:

- *Preparation*—Ensures that all the tools are working properly and are in the right location
- *Mounting and Extraction*—Involves the removal of the tooling after the production lot is completed and the placement of the new tooling before the next production lot
- *Establishing Control Settings*—Setting all the process control settings prior to the production run. Inclusive of calibrations and measurements needed to make the machine, tooling operate effectively
- *First Run Capability*—This includes the necessary adjustments (recalibrations, additional measurements) required after the first trial pieces are produced
- *Setup Improvement*—The time after processing during which the tooling, machinery is cleaned, identified, and tested for functionality prior to storage

TABLE 5.9

Proportion of Setup Times Before SMED Improvements

Setup Steps	Setup Type Traditional Internal	Setup Type Traditional External	Resource Consumption (%)	Setup Type One Step Internal	Setup Type One Step External
Preparation	X		20%		X
Mounting & Extraction	X		5%	X	
Establish Control Setting	X		15%		X
First Run Capability	X		50%	N/A	N/A
Process Improvement	X		10%		X

To determine the proportion of current setup times, the chart shown in Table 5.9 can be completed.

The three stages of SMED are explained next.

Description of Stage 1—Separate Internal vs. External Setup

Three techniques help us separate internal vs. external setup tasks:

1. Use checklists.
2. Perform function checks.
3. Improve transport of die and other parts.

Checklists

A checklist lists everything required to set up and run the next operation. The list includes items such as the following:

- Tools, specifications, and workers required.
- Proper values for operating conditions such as temperature, pressure, and so on.
- Correct measurement and dimensions required for each operation.
- Checking item of the list before the m/c is stopped helps prevent mistakes that come up after internal setup begun.

An operation checklist is shown in the Table 5.10.

TABLE 5.10

Operation Checklist

Operation Checklist
Equipment:
Operation:
Date:

Effective: Nov-11

	Employees Trained for Setup and Operations (Need 2 people)		
	Name of Employee	x	Name of Employee
x	Name of Employee		Name of Employee
	Tools Needed		
	Automatic Nut Driver		
x	Hex Wrench		
x	Rolling Cart		
	Tool		
	Tool		
	Tool		
	Tool		
x	Tool		
	Parts Needed		
x	Elevator Plate—3.5 lb. Size		
x	Compression Plate—3.5 lb. Size		
x	Feed Auger		
	Part		
	Part		
	Part		
	Part		
	Part		
	Standard Operating Procedure to Follow		
x	SOP001—Change over Procedure		
x	SOP003—Cleandown		
	Procedure		
	Procedure		
	Procedure		
	Procedure		
	Procedure		

Function Checks

- Should be performed before setup begins so that repair can be made if something does not work right.
- If broken dies, molds, or jigs are not discovered until test runs are done, a delay will occur in internal setup.
- Making sure such items are in working order before they are mounted will cut down setup time a great deal.

Improved Transport of Parts and Tools

- Dies, tools, jigs, gauges, and other items needed for an operation must be moved between storage areas and machines, then back to storage once a lot is finished.
- To shorten the time the machine is shut down, transport of these items should be done during external setup.
- In other words, new parts and tools should be transported to the machine before the machine is shutdown for changeover.

Description of Stage 2—Convert Internal Setups to External Setups

I. Advance Preparation of Conditions

- Get necessary parts tools and conditions ready before internal setup begins.
- Conditions like temperature, pressure, or position of material can be prepared externally while the machine is running (e.g., preheating of mold/material).

II. Function Standardization

- It would be expensive and wasteful to make external dimensions of every die, tool, or part the same, regardless of the size or shape of the product it forms. Function standardizations avoid this waste by focusing on standardizing only those elements whose functions are essential to the setup.
- Function standardization might apply to dimensioning, centering, securing, expelling, or gripping.

III. Implementing Function Standardization with Two Steps

- Look closely at each individual function in your setup process and decide which functions if any can be standardized.
- Look again at the functions and think about which can be made more efficient by replacing the fewest possible parts (e.g., clamping function standardization).

TABLE 5.11

Internal vs. External Setups Table

Classify items under each category.

	Internal		External
1		1	
2		2	
3		3	
4		4	
5		5	
6		6	
7		7	
8		8	
9		9	
10		10	

Which items would you convert from internal to external set up?

1		1	
2		2	
3		3	
4		4	
5		5	

Why?

1			
2			
3			
4			
5			

Internal versus external setups can be put in a table as in Table 5.11.

Description of Stage 3—Streamline All Aspects of the Setup Operation

- External setup improvement include streamlining the storage and transport of parts and tools.
- In dealing with small tools, dies, jigs, and gauges, it is vital to address issue of tool and die management.

Ask Questions

- What is the best way to organize these items?
- How can we keep these items maintained in perfect condition and ready for the next operation?
- How many of these items should we keep in stock?

Improving Storage and Transport

- Operation for storing and transporting dies can be very time consuming, especially when your factory keeps a large number of dies on hand.
- Storage and transport can be improved by marking the dies with color codes and location numbers of the shelves where they are stored.

Streamlining Internal Setup

- Implement parallel operations, using functional clamps, eliminating adjustments, and mechanization.

Implementing Parallel Operations

Machines such as plastic molding machines and die-casting machines often require operation at both the front and back of the machine. One-person changeovers of such machines mean wasted time and movement because the same person is constantly walking back and forth from one end of the machine to the other.

Parallel operations divide the setup operation between two people, one at each end of the machine. When setup is done using parallel operations, it is important to maintain reliable and safe operations and minimize waiting time. To help streamline parallel operations, workers should develop and follow procedural charts for each setup.

A Setup Conversion Matrix is shown in Table 5.12.

The final understanding of SMED comes from basic principles such as observing with videos. If there is nobody on the screen, it means there is waste present.

It is important to understand that SMED is more than just a series of techniques. It is a fundamental approach to improvement activities. A personal action plan should be found to adhere to each business's needs. It is important to find ways to implement SMED into environments to continue the sustainability of the businesses. To begin the process, a communication plan should be implemented.

TPM—Total Productive Maintenance

TPM has been a well-known activity that has several names associated within. Many people associate TPM with Total Predictive Maintenance or Total Preventative Maintenance. The association explained below will be Total Productive Maintenance but includes the above as well.

TABLE 5.12

Setup Conversion Matrix

Sheet			Date:	Page ____ of ____	
Area/Department	Machine/Equipment Name	Set-up Tools Required	Operator Number		Standard set-up Time Minutes
			Date Prepared		

Current Process		Current Time			Proposed Time	
NO.	**Task/Operation**	**Internal**	**External**	**Improvement**	**Internal**	**External**
	Current Total:			Improve Total		

Conversation Methology

- Preparation of Set-up Process
- Combining Equipment Functionality
- Standardized Jigs

TPM is performed in the improve phase based on downtimes or efficiency losses. The downtimes associated can be planned or unplanned. The goal of TPM is to increase all operational equipment efficiencies to above 85% by eliminating any wasted time such as setup time (see SMED section), idle times, downtimes, start-up delays, and any quality losses.

TPM ensures minimal downtime but in turn requires no defects as well. There are three basic steps for TPM that have several steps within each.

1. Analyze the current processes
   a. Calculate any costs associated with maintenance
   b. Calculate Overall Equipment Effectiveness (OEE) by finding the proportion of quality products produced at a given line speed

2. Restore equipment to its original and high operating states
   a. Inspect the machinery
   b. Clean the machinery
   c. Identify necessary repairs on the machinery
   d. Document defects
   e. Create a scheduling mechanism for maintenance
   f. Ensure maintenance has repaired machinery and improvements are sustained

3. Preventative maintenance to be carried out
   a. Create a schedule for maintenance with priorities—include high machinery defects, replacement parts, and any pertinent information
   b. Create stable operations—complete root cause analysis on high machinery defects and machinery that causes major downtime
   c. Create a planning and communication system—documentation of preventative maintenance activities should be accessible to all people so planning and prioritization within is completed
   d. Create processes for continuous maintenance—inspections should occur regularly and servicing for any machinery should be noted on a scheduled basis
   e. Internal operations should be optimized—any internal operations should be benchmarked with improvements from other areas to eliminate time spent on root cause analysis. When defects of machinery are not understood, it is important to put the machinery back to its original state to understand the root causes more efficiently. Time to exchange parts or retrieve parts should also be minimized.

f. Continuous improvement on preventative maintenance—Train employees for early detection of problems and maintenance measures. Visual controls should be put in place for changeovers. 5S should take place to eliminate wasted time. The documentation should be communicated and plans should be given regularly. All aspects should be looked upon to see if continuous improvements can be made.

The key TPM indicators will be able to show the following main issues:

- OEE
- Mean time between failures
- Mean time to repair

TPM is crucial to sustainability because it involves all the employees including high-level managers and creates planning for preventative maintenance so the issues are fixed before they become an error or defect. TPM also is a journey for educating and training the workforce to be familiar with machinery, parts, processes, and damages while being productive.

Design for Six Sigma

Design for Six Sigma (DFSS) is another process that is included in a phase called DMADV, which stands for the following:

Define
Measure
Analyze
Design
Verify

The difference between DMADV and DMAIC is the design and verification portions. DMAIC is process improvement driven whereas DMADV is for designing new products or services. Design stands for the designing of new processes required including the implementation. *Verify* stands for the results being verified and the performance of the design to be maintained.

The purpose of DFSS is very similar to the regular DMAIC cycle where it is a customer-driven design of processes with Six Sigma capabilities. DFSS does not only have to be manufacturing driven; the same methodologies can be used in service industries. The process is top-down with flow-down CTQs that match flow-up capabilities. DFSS is quality based where predictions are made regarding first-pass quality. The quality measurements are driven through predictability in the early design phases. Process capabilities are utilized to make final design decisions.

Finally, process variances are monitored to verify that Six Sigma customer requirements are met.

The main tools utilized in DFSS are FMEAs, Quality Function Deployment (QFD), Design of Experiments (DOE), and simulations.

Quality Function Deployment

Dr. Yoji Akao developed Quality Function Deployment (QFD) in 1966 in Japan. There was a combination of quality assurance and quality control that led to value engineering analyses. The method for QFD is simply to utilize consumer demands into designing quality functions and methods to achieve quality into subsystems and specific elements of processes. The basis for QFD is to take customer requirements from the Voice of the Customer and relay them into engineering terms to develop products or services. Graphs and matrices are utilized for QFD. A house type matrix is compiled to ensure that the customer needs are being met in the transformation of the processes or services designed. See Table 5.13.

The QFD house is a simple matrix where the legend is used to understand quality characteristics, customer requirements, and completion.

Design of Experiments

Design of Experiments (DOE) is an experimental design that shows what is useful, what is a negative connotation, and what has no effect. The majority of the time, 50% of the designs have no effect.

DOEs require a collection of data measurements, systematic manipulation of variables also known as factors placed in a prearranged way (experimental designs), and control for all other variables. The basis behind DOEs is to test everything in a prearranged combination and measure the effects of each of the interactions.

The following DOE terms are used:

- Factor: An independent variable that may affect a response
- Block: A factor used to account for variables that the experimenter wishes to avoid or separate out during analysis
- Treatment: Factor levels to be examined during experimentation
- Levels: Given treatment or setting for an input factor

TABLE 5.13
QFD House

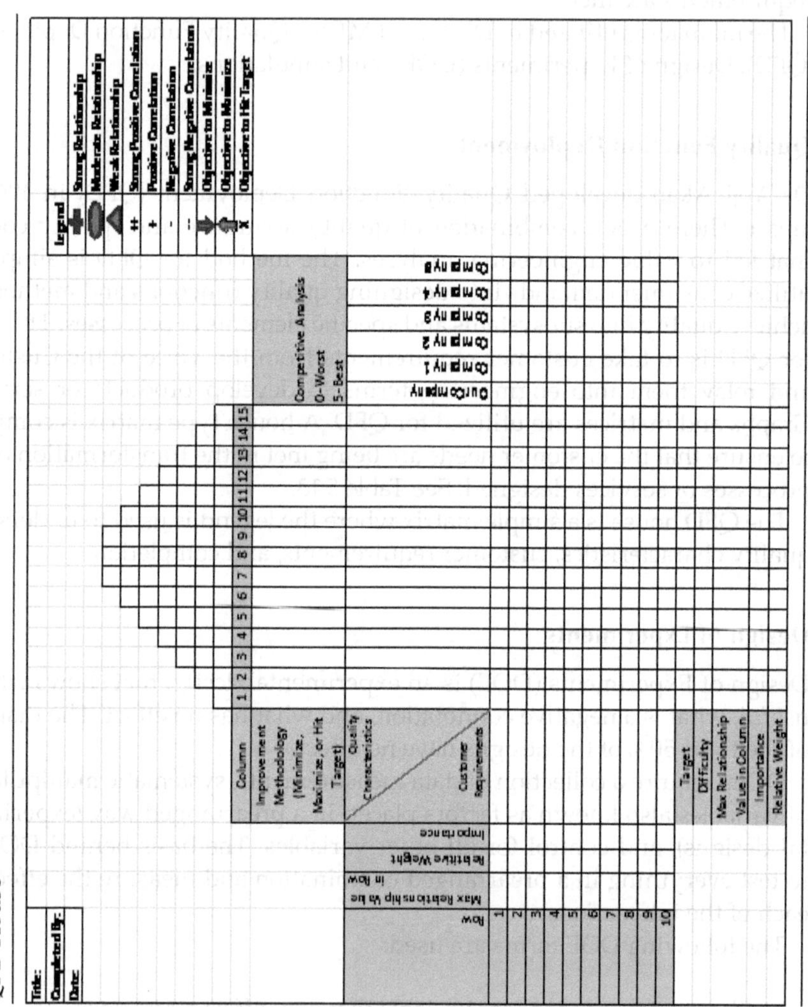

TABLE 5.14
Full-Factorial DOE

	Factors			Factor Interactions			
Number	**A**	**B**	**C**	**AB**	**AC**	**BC**	**ABC**
1	–	–	–	+	+	+	–
2		–	–	–	–	+	+
3	–	+	–	–	+	–	+
4	+	+	–	+	–	–	–
5	–	–	+	+	–	–	+
6	+	–	+	–	+	–	–
7	–	+	+	–	–	+	–
8	+	+	+	+	+	+	+

- Response: The result of a single run of an experiment at a given setting (or given combination of settings when more than one factor is involved)
- Replication (Replicate): Repeated run(s) of an experiment at a given setting (or given combination of settings when more than one factor is involved)

There are two types of DOEs: Full-Factorial Design and Fractional Factorial Design.

Full-Factorial DOEs determine the effect of the main factors and factor interactions by testing every factorial combination.

A Full-Factorial DOE factors all levels combined with one another covering all interactions. The basic design of a three factorial DOE is shown in Table 5.14.

The effects from the Full-Factorial DOE can then be calculated and sort into main effects and effects generated by interactions.

Effect = Mean Value of Response when Factor Setting is at High Level (Y_{A+}) – Mean Value of Response when Factor Setting is at Low Level (Y_{A-})

In a full-factorial experiment, all of the possible combinations of factors and levels are created and tested.

A two-level design (where each factor has two levels) with k factors, there are 2^k possible scenarios or treatments.

- 2 factors each with 2 levels, we have $2^2 = 4$ treatments
- 3 factors each with 2 levels, we have $2^3 = 8$ treatments
- k factors each with 2 levels, we have 2^k treatments

The analysis behind the DOE consists of the following steps:

1. Analyze the data.
2. Determine factors and interactions.
3. Remove statistically insignificant effects from the model such as p-values of less than 0.1 and repeat the process.
4. Analyze residuals to ensure the model is set correctly.
5. Analyze the significant interactions and main effects on graphs while setting up a mathematical model.
6. Translate the model into common solutions and make sustainable improvements.

A Fractional Factorial Design locates the relationship between influencing factors in a process and any resulting processes while minimizing the number of experiments. Fractional Factorial DOEs reduce the number of experiments while still ensuring the information lost is as minimal as possible. These types of DOEs are used to minimize time spent and money spent and to eliminate factors that seem unimportant.

The formula for a Fractional Factorial DOE is

$$2^{k-q}$$

where q equals the reduction factor.

The Fractional Factorial DOE requires the same number of positive and negative signs as a Full-Factorial DOE.

The Fractional Factorial DOE is shown in a matrix in Table 5.15.

TABLE 5.15

Fractional Factorial DOE

Factional Factorial Matrix												D				
Run	(1)	A	B	C	D	AB	AC	AD	BC	BD	CD	ABC	ABD	ACD	BCD	ABCD
1	+	−	−	−	−	+	+	+	+	+	+	−	−	−	−	+
2	+	+	−	−	+	−	−	+	+	−	−	+	−	−	+	+
3	+	−	+	−	+	−	+	−	−	+	−	+	−	+	−	+
4	+	+	+	−	−	+	−	−	−	−	+	−	−	+	+	+
5	+	−	−	+	+	+	−	−	−	−	+	+	+	−	−	+
6	+	+	−	+	−	−	+	−	−	+	−	−	+	−	+	+
7	+	−	+	+	−	−	−	+	+	−	−	−	+	+	−	+
8	+	+	+	+	+	+	+	+	+	+	+	+	+	+	+	+

Control Charts

Control charts are a great interpretation of understanding whether projects are being sustained by utilizing process monitoring. The process spread can be understood through control charts while also interpreting whether the process is in control and predictable. Common cause and special cause variation is able to be found through control charts. The amount of samples taken is an important aspect to control charts along with the frequency of sampling. It is important to have a random yet normal pattern of data. For example, if data is taken during normal operating conditions and then one data point is taken during a changeover, the data will be skewed and show a point out of control.

The following cheat sheet in Figure 5.27 can be used to determine the proper type of control chart to use.

X-Bar and Range Charts

The *R*-chart is a time plot useful for monitoring short-term process variations. The *X*-bar chart monitors longer-term variations where the likelihood of special causes is greater over time. Both charts utilize control lines called upper and lower control limits and central lines; both types of lines are calculated from process measurements. They are not specification limits or percentages of the specifications or other arbitrary lines based on experience. They represent what a process is capable of doing when only common cause variation exists, in which case the data will continue to fall in a random fashion within the control limits and the process is in a state of statistical control. However, if a special cause acts on a process, one or more data points will be outside the control limits and the process will no longer be in a state of statistical control.

The following components should be used for control chart purposes:

- UCL—Upper Control Limit
- LCL—Lower Control Limit
- CL—Center Line; shows where the characteristic average falls
- USL—Upper specification limit or upper customer requirement
- LSL—Lower specification limit or lower customer requirement

Control limits describe the stability of the process. Specifically, control limits identify the expected limits of normal, random, or chance variation that is present in the process being monitored. Control limits are set by the process.

Specification limits are those limits that describe the characteristics the product or process must have to conform to customer requirements or to perform properly in the next operation.

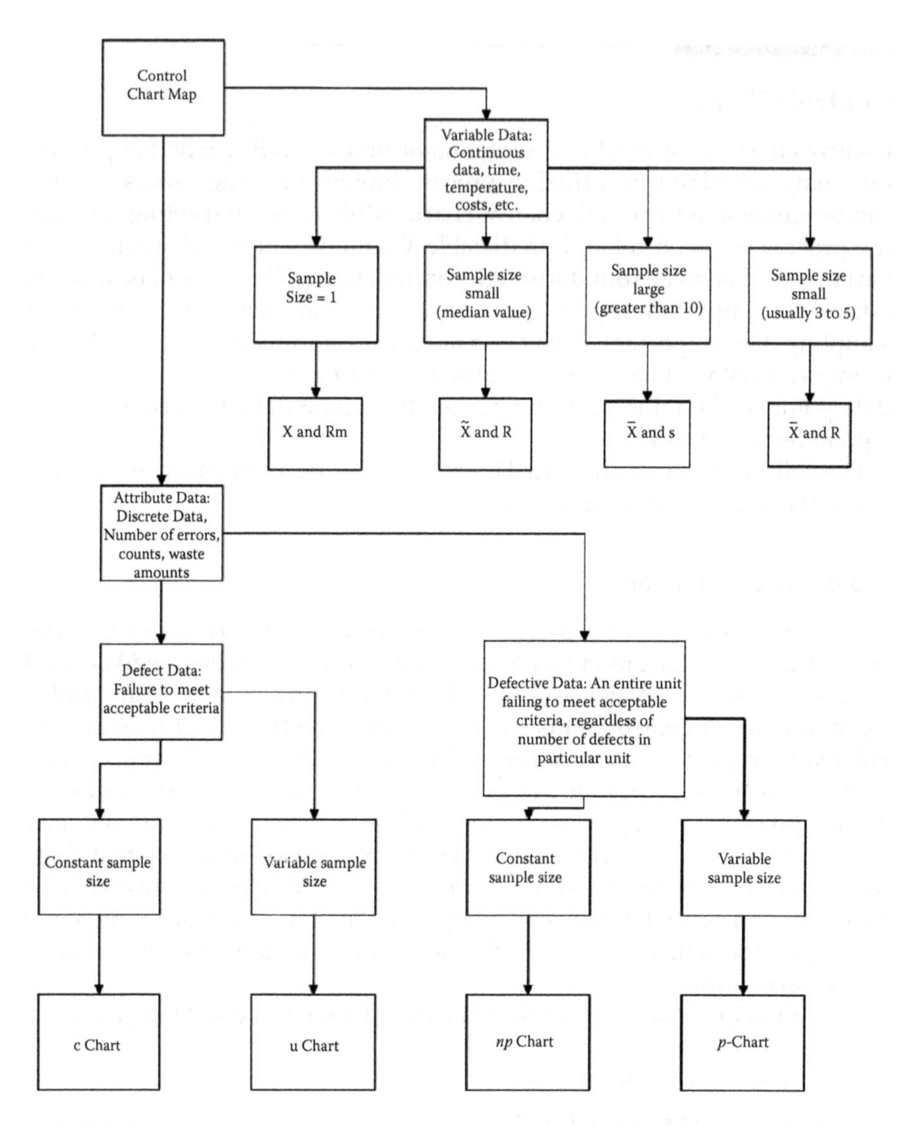

FIGURE 5.27
Control chart map.

The following Shewhart constants listed in Table 5.16 should be used when applicable to the formulas.

There are nine steps to constructing $\overline{X}$ and R-charts:

1. Identify characteristic, measurement method, and sampling pattern.
2. Record data (time).
3. Calculate sample average, and sample range R.
4. Calculate grand average and average range $\overline{\overline{X}}$ and $\overline{R}$.

TABLE 5.16

Shewhart Constants

n	A2	A3	d2	D3	D4	B3	B4
2	1.880	2.659	1.128	0.000	3.267	0.000	3.267
3	1.023	1.954	1.693	0.000	2.574	0.000	2.568
4	0.729	1.628	2.059	0.000	2.282	0.000	2.266
5	0.577	1.427	2.326	0.000	2.114	0.000	2.089
6	0.483	1.287	2.534	0.000	2.004	0.030	1.970
7	0.419	1.182	2.704	0.076	1.924	0.118	1.882
8	0.373	1.099	2.847	0.136	1.864	0.185	1.815
9	0.337	1.032	2.970	0.184	1.816	0.239	1.761
10	0.308	0.975	3.078	0.223	1.777	0.284	1.716
11	0.285	0.927	3.173	0.256	1.744	0.321	1.679
12	0.266	0.886	3.258	0.283	1.717	0.354	1.646
13	0.249	0.850	3.336	0.307	1.693	0.382	1.618
14	0.235	0.817	3.407	0.328	1.672	0.406	1.594
15	0.223	0.789	3.472	0.347	1.653	0.428	1.572
16	0.212	0.763	3.532	0.363	1.637	0.448	1.552
17	0.203	0.739	0.588	0.378	1.622	0.466	1.534
18	0.194	0.718	3.640	0.391	1.608	0.482	1.518
19	0.187	0.698	3.689	0.403	1.597	0.497	1.503
20	0.180	0.680	3.735	0.415	1.585	0.510	1.490
21	0.173	0.663	3.778	0.425	1.575	0.523	1.477
22	0.167	0.647	3.819	0.434	1.566	0.534	1.466
23	0.162	0.633	3.858	0.443	1.557	0.545	1.455
24	0.157	0.619	3.895	0.451	1.548	0.555	1.445
25	0.153	0.606	3.931	0.459	1.541	0.565	1.435

5. If stable, calculate limits.
6. Calculate control limits:

$$UCL\bar{x} = \bar{\bar{x}} + A_2\bar{R}$$

$$LCL\bar{x} = \bar{\bar{x}} - A_2\bar{R}$$

$$UCLR = D_4\bar{R}$$

$$LCLR = D_3\bar{R}$$

7. Construct control charts.
8. Plot initial data points.
9. Interpret chart with respect to variation: common cause variation or special cause variation.

Attribute Data Formulas

The following formulas should be used for attribute data when constructing the particular control charts:

p-chart:

$$p = \frac{\text{number of defective units}}{\text{number of units inspected}}$$

$$\bar{p} = \frac{\text{number of defectives in all samples}}{\text{number of units in all samples}} = \frac{\sum p_j n_j}{\sum n_j}$$

$$UCL = \bar{p} + \frac{3\sqrt{\bar{p}(1-\bar{p})}}{\sqrt{\bar{n}}}$$

$$LCL = \bar{p} - \frac{3\sqrt{\bar{p}(1-\bar{p})}}{\sqrt{\bar{n}}}$$

np-chart:

$$UCL = \bar{n}\,\bar{p} + 3\sqrt{\bar{n}\,\bar{p}(1-\bar{p})}$$

$$LCL = \bar{n}\,\bar{p} - 3\sqrt{\bar{n}\,\bar{p}(1-\bar{p})}$$

C-chart:

$$\bar{c} = \sum u_j \qquad UCL = \bar{c} + 3\sqrt{\bar{c}}$$

$$\sum n_j \qquad LCL = \bar{c} - 3\sqrt{\bar{c}}$$

where u_j = number of defects in the *j*th sample

U-chart:

$$\bar{u} = \frac{\text{number of defects in all samples}}{\text{number of units in all samples}} = \frac{\sum u_j}{\sum n_j}$$

$$UCL = \bar{u} + \frac{3\sqrt{\bar{u}}}{\sqrt{\bar{n}}}$$

TABLE 5.17

Number Defects per Month

Month	Errors	Inspected	Proportion
January	56	100	0.56
February	59	87	0.67816092
March	69	90	0.766666667
April	72	94	0.765957447
May	44	80	0.55
June	39	110	0.354545455
July	46	88	0.522727273
August	25	76	0.328947368
September	20	82	0.243902439
October	15	80	0.1875
November	21	77	0.272727273
December	18	66	0.272727273

The following rules should be used to determine if a process is out of control for both attribute and variable data:

1. One or more points fall outside of the control limits.
2. Two points out of three consecutive points are on the same side of the average.
3. Four points out of five points are on the same side of the average.
4. Nine consecutive points are on one side of the average.
5. Six consecutive points are increasing or decreasing.
6. Fourteen consecutive points trend upward or downward.
7. Fifteen consecutive points are above or below the average.

An example of how to utilize a control chart to represent if a process is sustainable after improvements would be to map out the number of defects per month (see Table 5.17). A *p*-chart could be useful in this case to see the proportion of defective units per sample produced due to a particular type of errors.

Example

The manufacturing process was unstable, having many defects per month. It was found out that the manufacturing equipment was not as reliable as possible, causing breaks in the system. The defects were monitored per month. A Six Sigma project was identified to improve the reliability of the equipment. The defects were mapped against the number of units inspected to see if the improvements that were implemented from August through December made an impact.

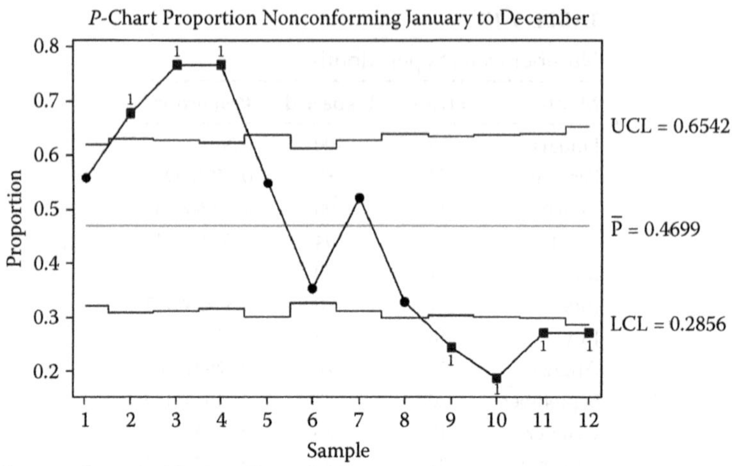

FIGURE 5.28
P-chart of defects.

A *p*-chart was mapped to see the improvements visually (see Figure 5.28). It is clear that there was a mean shift over the last 4 months that shows a decrease in errors. More points would need to be taken to prove the validity of the improvements over time. The conclusion would be that there was an improvement.

Control limits describe the representative nature of a stable process. Specifically, control limits identify the expected limits of normal, random, or chance variation that is present in the process being monitored. Sustainable processes must follow these rules.

Control Plans

A control plan is a vital part of sustainability because without it, there is no sustainability. A control plan takes the improvements made and ensures that they are being maintained and continuous improvement is achieved. A control plan is a very detailed document that includes who, what, where, when, and why (the why is based on the root cause analysis). The 12 basic steps of a control plan are listed below:

1. Collect existing documentation for the process.
2. Determine the scope of the process for the current control plan.

3. Form teams to update the control plan regularly.
4. Replace short-term capability studies with long-term capability results.
5. Complete control plan summaries.
6. Identify missing or inadequate components or gaps.
7. Review training, maintenance, and operational action plans.
8. Assign tasks to team members.
9. Verify compliance of actual procedures with documented procedures.
10. Retrain operators.
11. Collect sign-offs from all departments.
12. Verify effectiveness with long-term capabilities.

A control plan ensures consistency while eliminating as much variation from the system as possible. The plan is essential to operators because it enforces SOPs and eliminates changes in processes. It also ensures that PMs are performed and the changes made to the processes are actually improving the problem that was found through the root cause analysis. Control plans hold people accountable if reviewed at least quarterly. A sample control plan is shown in Figure 5.29.

Six Sigma Control Plan

Product:				Core Team:					Date (orig):		
Key Contact:									Date (revised):		
Phone:											
Process	Process Step	Input	Output	Process Specs (LSL, USL, Target)	Pkg./Date	Measurement Technique	%P/T	Sample Size	Sample Frequency	Control Method	Reaction Plan

FIGURE 5.29
Control plan.

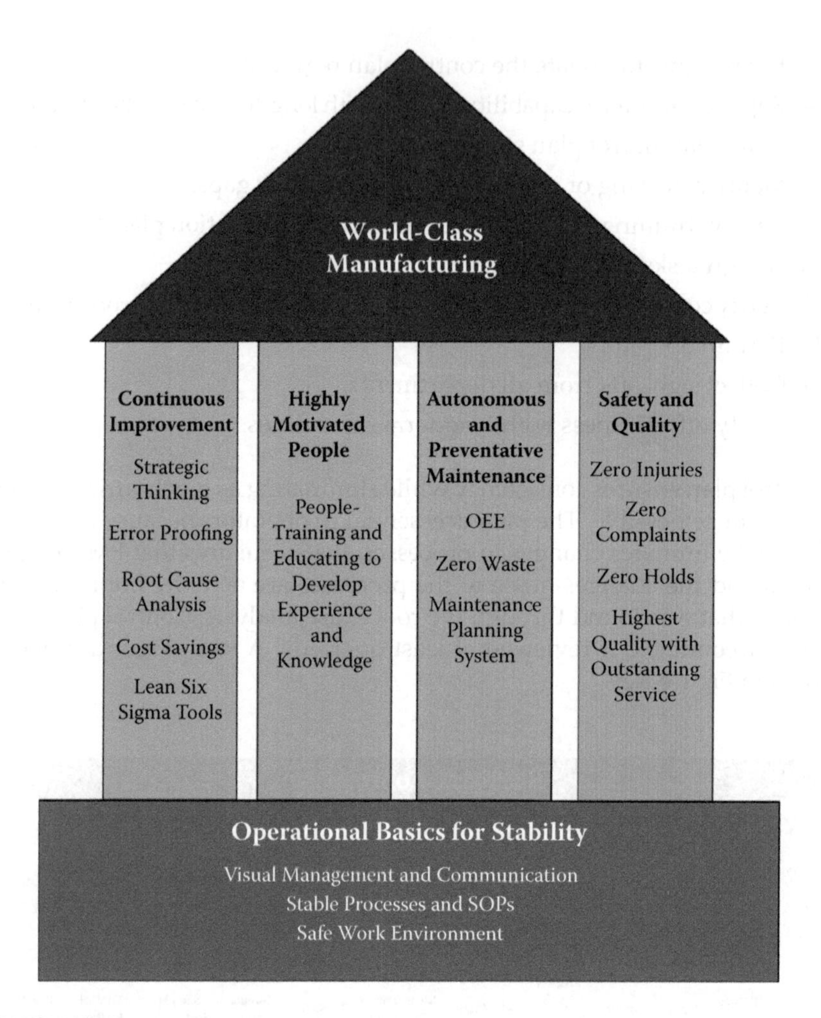

FIGURE 5.30
World-class manufacturing house.

Finally, a house is created for operational stability and sustainability that encompass the earlier discussed topics shown in Figure 5.30. The house can be rearranged or reworded with different goals and tools set out for the particular business or manufacturing environment.

The base of the house shows the stability and the areas needed for a stable and sustainable work environment. The pillars are the goals and tools utilized within the house in order for the roof to be stable upon the pillars. The roof of the house is the ultimate goal for a beneficial environment.

References

Fehlmann, T. M. (2011). "Do We Need a Matrix for QFD?" QFD Online. Accessed from http://www.qfdonline.com/templates/3f2504e0-4f89-11d3-9a0c-0305e82c2899/.

George, M., Rowlands, D., Price, M., and Maxey, J. (2005). *The Lean Six Sigma Pocket Toolbook*. McGraw-Hill, New York.

Kovach, T. (2012). *Statistical Techniques for Project Control*. Taylor & Francis, New York.

Lunau, S., John, A., Roenpage, O., and Staudter, C. (2008). *Six Sigma + Lean Toolset*. Springer-Verlag, Berlin.

The Productivity Press Development Team. (2010). *Mistake-Proofing for Operators: The ZQC System*. Taylor & Francis, New York.

Theory of Constraints. Accessed from http://www.bexcellence.org/Theory-of-Constraints.html.

6

Technology Transfer for Sustainability

Why reinvent the wheel when you can ride the bus?

—**Adedeji Badiru (1993)**

No single entity has all the answers. To achieve sustainability, we must, at one point or another, rely on technology transfer from one point to another. The focus of technology transfer can be in terms of a physical product, a service element, a best-practice result, or even a motivating concept (Badiru, 1993).

The concepts of project management can be very helpful in planning for the adoption and implementation of new sustainability technology. Due to its many interfaces, the area of sustainability technology adoption and implementation is a prime candidate for the application of project planning and control techniques. Technology managers, engineers, and analysts should make an effort to take advantage of the effectiveness of project management tools. This applies the various project management techniques that have been discussed in the preceding chapters to the problem of sustainability technology transfer. Project management approach is presented within the context of technology adoption and implementation for sustainability development.

Project management guidelines are presented for sustainability technology management. The Triple C model of Communication, Cooperation, and Coordination is applied as an effective tool for ensuring the acceptance of new technology. The importance of new technologies in improving product quality and operational productivity is also discussed.

Definition and Characteristics of Technology

To transfer technology, we must know what constitutes technology. A working definition of technology will enable us to determine how best to transfer it. A basic question that should be asked is: What is technology? How does it relate to sustainability? Technology can be defined as follows:

> Technology is a combination of physical and nonphysical processes that make use of the latest available knowledge to achieve business, service, or production goals.

Technology is a specialized body of knowledge that can be applied to achieve a mission or a purpose. For our own case, sustainability is the object of our interest. The knowledge concerned could be in the form of methods, processes, techniques, tools, machines, materials, and procedures. Technology design, development, and effective use is driven by effective utilization of human resources and effective management systems. Technological progress is the result obtained when the provision of technology is used in an effective and efficient manner to improve productivity, reduce waste, improve human satisfaction, and raise the quality of life.

Technology all by itself is useless. However, when the right technology is put to the right application, with effective supporting management system, it can be very effective in achieving sustainability goals. Technology implementation starts with an idea and ends with a productive sustainability process. Technological progress is said to have occurred when the outputs of technology, in the form of information, instrument, or knowledge, leads to a lowering of costs of production, better product quality, and higher levels of output (from the same amount of inputs). The information and knowledge involved in technological progress include those which improve the performance of management, labor, and the total resources expended for a given activity.

Technological progress plays a vital role in improving overall national productivity. Experience in developed countries such as in the United States show that in the period 1870–1957, 90% of the rise in real output per man-hour can be attributed to technological progress. Industrial or economic growth is dependent on improvements in technical capabilities as well as on increases in the amount of the conventional factors of capital and labor. It is noted that technical change is not synonymous with a movement toward the mostly modern capital-intensive process. Changes occur through improvements in the efficiency in use of existing equipment—that is, through learning and through the adaptation of other technologies, some of which may involve different collections of equipment. The present challenge is how to develop the infrastructure that promotes, uses, and develops technological knowledge to benefit the goals of sustainability.

Most of the developing nations today face serious challenges arising not only from the worldwide imbalance of dwindling revenue from industrial products but also from major changes in a world economy that is characterized by competition, imports, and exports of not only oil but also basic technology, weapon systems, and electronics products. If technology utilization is not given the right attention in all sectors of the national economy, the much-desired sustainability development cannot take place. The ability of a nation to compete in the world market will, consequently, be stymied.

The important characteristics or attributes of a new technology may include productivity improvement, improved product quality, production cost savings, flexibility, reliability, and safety. An integrated evaluation must be performed to ensure that a proposed technology is justified both economically and technically. The scope and goals of the proposed technology must be established right from the beginning of the project. This entails the comparison of departmental objectives with the overall organizational goals in the following areas:

1. Industrial marketing strategy: This should identify the customers of the proposed technology. It should also address items such as market cost of proposed product, assessment of competition, and market share. Import and export considerations should be a key component of the marketing strategy.
2. Industry growth and long-range expectations: This should address short-range expectations, long-range expectations, future competitiveness, future capability, and prevailing size and strength of the industry that will use the proposed technology.
3. National benefit: Any prospective technology must be evaluated in terms of direct and indirect benefits to be generated by the technology. These may include product price versus value, increase in international trade, improved standard of living, cleaner environment, safer workplace, and improved productivity.
4. Economic feasibility: An analysis of how the technology will contribute to profitability should consider past performance of the technology, incremental benefits of the new technology versus conventional technology, and value added by the new technology.
5. Capital investment: Comprehensive economic analysis should play a significant role in the technology assessment process. This may cover an evaluation of fixed and sunk costs, cost of obsolescence, maintenance requirements, recurring costs, installation cost, space requirement cost, capital substitution potentials, return on investment, tax implications, cost of capital, and other concurrent projects.
6. Resource requirements: The utilization of resources (manpower and equipment) in the pre-technology and post-technology phases of industrialization should be assessed. This may be based on material input/output flows, high value of equipment versus productivity improvement, required inputs for the technology, expected output of the technology, and utilization of technical and nontechnical personnel.
7. Technology stability: Uncertainty is a reality in technology adoption efforts. Uncertainty will need to be assessed for the initial investment, return on investment, payback period, public reactions, environmental impact, and volatility of the technology.
8. National productivity improvement: An analysis of how the technology may contribute to national productivity may be verified by studying industry throughput, efficiency of production processes, utilization of raw materials, equipment maintenance, absenteeism, learning rate, and design-to-production cycle.

Technology Assessment

New service technologies have been gaining more attention in recent years. This is due to the high rate at which new productivity improvement technologies are being developed. The fast pace of new technologies has created difficult implementation and management problems for many organizations. New technology can be successfully implemented only if it is viewed as a system whose various components must be evaluated within an integrated managerial framework. Such a framework is provided by a project management approach.

It is important to consider the peculiar characteristics of a new technology before establishing adoption and implementation strategies. The justification for the adoption of a new technology is usually a combination of several factors rather than a single characteristic of the technology. The potential of a specific technology to contribute to sustainability development goals must be carefully assessed. The technology assessment process should explicitly address the following questions:

What is expected from the new technology?

Where and when will the new technology be used?

How is the new technology similar to existing technologies?

How is the proposed technology different from existing technologies?

What is the availability of technical personnel to support the new technology?

What is the administrative support for the new technology?

Who will use the technology, and when and how will it be used?

The development, transfer, adoption, utilization, and management of technology are problems that are faced in one form or another by business, industry, and government establishments. Some of the specific problems in technology transfer and management include the following:

- Controlling technological change
- Integrating technology objectives
- Shortening the technology transfer time
- Identifying a suitable target for technology transfer
- Coordinating the research and implementation interface
- Formal assessment of current and proposed technologies
- Developing accurate performance measures for technology

- Determining the scope or boundary of technology transfer
- Managing the process of entering or exiting a technology
- Understanding the specific capability of a chosen technology
- Estimating the risk and capital requirements of a technology

Integrated managerial efforts should be directed at the solution of the problems stated above. A managerial revolution is needed to cope with the ongoing technological revolution. The revolution can be instituted by modernizing the long-standing and obsolete management culture relating to technology transfer. Some of the managerial functions that will need to be addressed when developing technology transfer strategies include the following:

1. Development of a technology transfer plan
2. Assessment of technological risk
3. Assignment/reassignment of personnel to effect the technology transfer
4. Establishment of a transfer manager and a technology transfer office. In many cases, transfer failures occur because no individual has been given the responsibility to ensure the success of technology transfer.
5. Identification and allocation of the resources required for technology transfer
6. Setting of guidelines for technology transfer. For example: specification of phases (Development, Testing, Transfer, etc.); specification of requirements for inter-phase coordination; identification of training requirements; establishment and implementation of performance measurement.
7. Identification of key factors (both qualitative and quantitative) associated with technology transfer and management
8. Investigation of how the factors interact and develop the hierarchy of importance for the factors
9. Formulation of a loop system model that considers the forward and backward chains of actions needed to effectively transfer and manage a given technology
10. Tracking of the outcome of the technology transfer

Technology development, in many industries, appears in scattered, narrow, and isolated areas within a few selected fields. This means that technology efforts are rarely coordinated, thereby hampering the benefits of technology. The optimization of technology utilization is, thus, very difficult. To overcome this problem and establish the basis for effective technology transfer and management, an integrated approach must be followed. An integrated approach will be applicable to technology transfer between any two organizations, public or private.

Some nations concentrate on the acquisition of bigger, better, and faster technology. But little attention is given to how to manage and coordinate the operations of the technology once it arrives. When technology fails, it is not necessarily because the technology is deficient. Rather, it is often the communication, cooperation, and coordination functions of technology management that are deficient. Technology encompasses factors and attributes beyond mere hardware and software. Consequently, technology transfer involves more than the physical transfer of hardware and software. Several flaws exist in the present practices involving technology transfer and management. These flaws include the following:

- Poor fit: Inadequate assessment of the need of the organization receiving the technology. The target of the transfer may not have the capability to properly absorb the technology.
- Premature transfer: This is particularly acute for emerging technologies that are prone to frequent developmental changes.
- Lack of focus: In the attempt to get a bigger share of the market or gain early lead in the technological race, organizations frequently force technology in many incompatible directions.
- Intractable implementation problems: Once a new technology is in place, it may be difficult to locate sources of problems that have their roots in the technology transfer phase.
- Lack of transfer precedents: Very few precedents are available on the management of new sustainability technology. Managers are, thus, often unprepared for sustainability management responsibilities.
- Unwillingness or inability to reorganize priorities: Unworkable technologies sometimes continue to be recycled needlessly in the attempt to find the "right" usage.
- Lack of foresight: Due to the nonexistence of a technology transfer model, managers may not have a basis against which they can evaluate future expectations.
- Insensitivity to external events: Some external events that may affect the success of sustainability technology transfer include trade barriers, tariffs, political changes, etc.
- Improper allocation of resources: There are usually not enough resources available to allocate to technology alternatives. Thus, a technology transfer priority must be developed.

The steps presented below provide a specific guideline for pursuing the implementation of sustainability technology.

1. Find a suitable application.
2. Commit to an appropriate technology.
3. Perform economic justification.
4. Secure management support for the chosen technology.
5. Design the technology implementation to be compatible with existing operations.
6. Formulate project management approach to be used.
7. Prepare the receiving organization for the technology change.
8. Install the technology.
9. Maintain the technology.
10. Periodically review the performance of the technology based on prevailing goals.

Sustainability Technology Transfer Modes

Transfer of technology can be achieved in various forms. Project management provides an effective means of ensuring proper transfer of technology. Three technology transfer modes are presented here to illustrate basic strategies for getting one technological product from one point (technology source) to another point (technology sink). A conceptual integrated model of the interaction between the technology source and sink is presented in Figure 6.1.

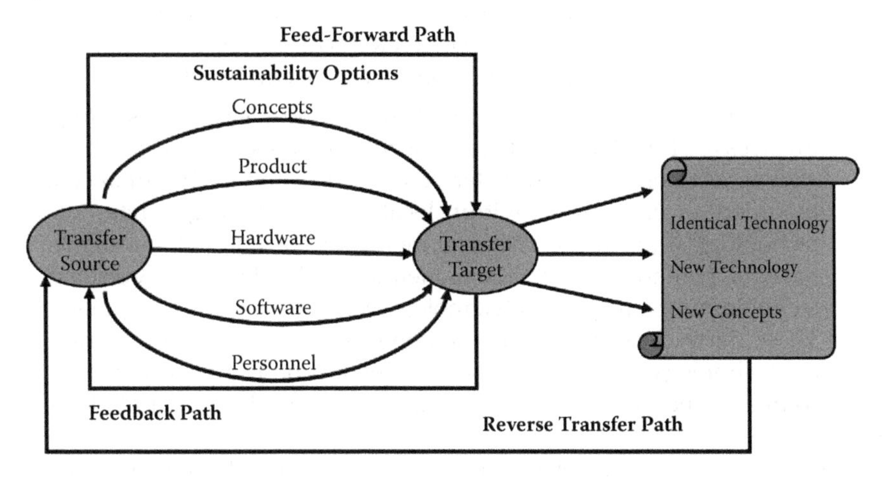

FIGURE 6.1
Technology transfer modes. (Adapted from Badiru, A.B., *Managing Industrial Development Projects: A Project Management Approach*, John Wiley & Sons, New York, 1993.)

The university–industry interaction model presented in Chapter 1 can be used as an effective mechanism for facilitating technology transfer. Sustainability technology application centers may be established to serve as a unified point for linking technology sources with desirous targets. The center will facilitate interactions between business establishments, academic institutions, and government agencies to identify important technology needs. Technology can be transferred in one or a combination of the following strategies:

1. Transfer of complete technological products. In this case, a fully developed product is transferred from a source to a target. Very little product development effort is carried out at the receiving point. However, information about the operations of the product is fed back to the source so that necessary product enhancements can be pursued. So, the technology recipient generates product information, which facilitates further improvement at the technology source. This is the easiest mode of technology transfer and the most tempting. Developing nations are particularly prone to this type of transfer. Care must be exercised to ensure that this type of technology transfer does not degenerate into "machine transfer." It should be recognized that machines alone do not constitute technology.
2. Transfer of technology procedures and guidelines. In this technology transfer mode, procedures (e.g., blueprints) and guidelines are transferred from a source to a target. The technology blueprints are implemented locally to generate the desired services and products. The use of local raw materials and manpower is encouraged for the local production. Under this mode, the implementation of the transferred technology procedures can generate new operating procedures that can be fed back to enhance the original technology. With this symbiotic arrangement, a loop system is created whereby both the transferring and the receiving organizations derive useful benefits.
3. Transfer of technology concepts, theories, and ideas. This strategy involves the transfer of the basic concepts, theories, and ideas behind a given technology. The transferred elements can then be enhanced, modified, or customized within local constraints to generate new technological products. The local modifications and enhancements have the potential to generate an identical technology, a new related technology, or a new set of technology concepts, theories, and ideas. These derived products may then be transferred back to the original technology source. An academic institution is a good potential source for the transfer of sustainability concepts, theories, and ideas.

It is very important to determine the mode in which technology will be transferred for sustainability development purposes. There must be a concerted effort by people to make transferred technology work within local situations. Local innovation, patriotism, dedication, and willingness to adapt technology will be required to make technology transfer successful. It will be difficult for a nation or an organization to achieve sustainability development through total dependence on transplanted technology. Local adaptation will always be necessary.

Sustainability Changeover Strategies

Any development project will require changing from one form of technology to another. The implementation of a new technology to replace an existing (or a nonexistent) technology can be approached through one of several options. Some options are more suitable than others for certain types of technologies. The most commonly used technology changeover strategies include the following:

1. Parallel changeover: The existing technology and the new technology operate concurrently until there is confidence that the new technology is satisfactory.
2. Direct changeover: The old technology is removed totally and the new technology takes over. This method is recommended only when there is no existing technology or when both technologies cannot be kept operational due to incompatibility or cost considerations.
3. Phased changeover: Modules of the new technology are gradually introduced one at a time using either direct or parallel changeover.
4. Pilot changeover: The new technology is fully implemented on a pilot basis in a selected department within the organization.

Post-Implementation Evaluation

The new technology should be evaluated only after it has reached a steady-state performance level. This helps to avoid the bias that may be present at the transient stage due to personnel anxiety, lack of experience, or resistance to change. The system should be evaluated for the following:

- Sensitivity to data errors
- Quality and productivity
- Utilization level

- Response time
- Effectiveness

Technology Systems Integration

With the increasing shortages of resources, more emphasis should be placed on the sharing of resources. Technology resource sharing can involve physical equipment, facilities, technical information, ideas, and related items. The integration of technologies facilitates the sharing of resources. Technology integration is a major effort in technology adoption and implementation. Technology integration is required for proper product coordination. Integration facilitates the coordination of diverse technical and managerial efforts to enhance organizational functions, reduce cost, improve productivity, and increase the utilization of resources. Technology integration ensures that all performance goals are satisfied with a minimum of expenditure of time and resources. It may require the adjustment of functions to permit sharing of resources, development of new policies to accommodate product integration, or realignment of managerial responsibilities. It can affect both hardware and software components of an organization. Important factors in technology integration include the following:

- Unique characteristics of each component in the integrated technologies
- Relative priorities of each component in the integrated technologies
- How the components complement one another
- Physical and data interfaces between the components
- Internal and external factors that may influence the integrated technologies
- How the performance of the integrated system will be measured

Sustainability Performance Evaluation

Time, cost, and performance form the basis for the operating characteristics of sustainability technology. These factors help to determine the basis for sustainability planning and control. Technology control is the process of reducing the deviation between actual performance and expected performance. To be able to control a technology, we must be able to measure its performance. Measurements are taken regarding time (schedule), performance, and cost.

The traditional procedures for measuring progress, evaluating performance, and taking control actions are not adequate for technology management where events are more dynamic. Some of the causes of sustainability technology control problems are summarized below.

Sustainability Schedule Problems

- Procrastination of technology adoption
- Poor precedence relationships
- Unreliable feasibility study
- Delay of critical activities
- Hasty implementation
- Technical problems
- Poor timing

Sustainability Performance Problems

- Inappropriate application
- Poor quality of hardware
- Lack of clear objectives
- Maintenance problems
- Poor functionality
- Improper location
- Lack of training

Sustainability Cost Problems

- Lack of complete economic analysis
- Inadequate starting budget
- Effects of inflation
- Poor cost reporting
- High overhead cost
- Unreasonable scope
- High labor cost

Sustainability Planning

Sustainability planning involves establishing the set of actions needed to achieve the goals of sustainability. Sustainability technology planning is needed to accomplish the following:

- Minimize the effects of technology uncertainties
- Clarify technology goals and objectives
- Provide basis for evaluating the progress of technology project
- Establish measures of technology performance
- Determine required personnel responsibilities

Technology Overview

This specifies the goals and scope of the technology as well as its relevance to the overall mission of the development project. The major milestones, with a description of the significance of each, should be documented. In addition, the organization structure to be used for the project should be established.

Technology Goal

This consists of a detailed description of the overall goal of the proposed technology. A technology goal may be a combination of a series of objectives. Each objective should be detailed with respect to its impact on the project goal. The major actions that will be taken to ensure the achievement of the objectives should also be identified.

Strategic Planning

The overall long-range purposes of the technology should be defined. Technologically feasible useful life should be defined. A frequent problem with technology is the extension of useful life well beyond the time of obsolescence. If a technologically feasible life is defined during the planning stage, it will be less traumatic to replace the technology at the appropriate time.

Sustainability Policy

Technology policy refers to the general guideline for personnel actions and managerial decision making relating to the adoption and implementation of new technology. The project policy indicates how the project plan will be executed. The chain of command and the network of information flow are governed by the established policy for the project. A lack of policy creates a fertile ground for incoherence in technology implementation due to conflicting interpretations of the project plan.

Sustainability Technology Procedures

Technology procedures are the detailed methods of complying with established technology policies. A policy, for example, may stipulate that the approval of the project manager must be obtained for all purchases.

A procedure then may specify how the approval should be obtained: oral or written. A policy without procedures creates an opportunity for misinterpretations.

Sustainability Resources

The resources (manpower and equipment) required for the adoption and implementation of new technology should be defined. Currently available technology resources should be identified along with resources yet to be acquired. The time frame of availability of each resource should also be specified. Issues such as personnel recruiting and technical training should be addressed early in the project.

Sustainability Technology Budget

Technology cannot be acquired without adequate budget. Some of the cost aspects that will influence technology budgeting are first cost, fixed cost, operating cost, maintenance cost, direct/indirect costs, overhead cost, and salvage value.

Sustainability Operating Characteristics

A specification of the operating characteristics of the technology should be developed. Questions about operating characteristics should include the following: What inputs will be required by the technology? What outputs are expected from the technology? What is the scope of the technology implementation? How will its performance be measured and evaluated? Is the infrastructure suitable for the technology's physical configuration? What maintenance is needed and how will the maintenance be performed? What infrastructure is required to support the technology?

Sustainability Cost/Benefit Analysis

The bottom line in any technology implementation is composed of profit, benefit, and/or performance. An analysis of the expenditure required for implementing the technology versus its benefits should be conducted to see if the technology is worthwhile. Can an existing technology satisfy the needs more economically? Even if future needs dictate the acquisition of the technology, economic decisions should still consider prevailing circumstances.

Technology Performance Measures

Performance standards should be established for any new technology. The standards provide the yardsticks against which adoption and implementation progress may be compared. In addition, the methods by which the performance will be analyzed should also be defined to avoid ambiguities in tracking and reporting.

Sustainability Technology Organization

Technology organization involves organizing the technology personnel with respect to required duties, assigned responsibilities, and desired personnel interactions. The organization structure serves as the coordination model for the technology implementation project.

Sustainability Work Breakdown Structure

This refers to a logical breakdown of the technology implementation project into major functional clusters. This facilitates a more efficient and logical analysis of the elements and activities involved in the adoption and implementation process. A work breakdown structure (WBS) shows the hierarchy of major tasks required to accomplish project objectives. It permits the implementation of the "divide and conquer" concepts. Overall technology planning and control can be improved by using WBS. A large project may be broken down into smaller subprojects, which may, in turn, be broken down into task groups.

Potential Technology Problems

New technologies are prone to new and unknown problems. Contingency plans must be established. Preparation must be made for unexpected problems such as technical failure, software bugs, personnel problems, technological changes, equipment failures, human errors, data deficiency, decision uncertainties, and so on.

Sustainability Technology Acquisition Process

Technology acquisition deals with the process of procuring and implementing a proposed technology. The acquisition may involve the acquisition of both physical and intangible assets. The acquisition process should normally cover the following analysis:

Hardware: This involves an analysis of the physical component of the technology. Questions that should be asked may concern such factors as size, weight, safety features, space requirement, and ergonomics.

Software: This relates to the analysis of the program code, user interface, and operating characteristics of any computer software needed to support the proposed technology.

Site selection and installation: A suitable and accessible location should be found for the physical component of the new technology. The surrounding infrastructure should be such that the function of the technology is facilitated.

A successful execution of technology projects requires a coordinated approach that should utilize conventional project planning and control techniques as well as other management strategies. Intricate organizational and human factors considerations come into play in the implementation of today's complex technologies for sustainability. The success of projects depends on good levels of communication, cooperation, and coordination. The Triple C model, which is addressed in a later chapter, can facilitate a systematic approach to the planning, organizing, scheduling, and control of a sustainability technology transfer endeavors.

Reference

Badiru, A.B. (1993). *Managing Industrial Development Projects: A Project Management Approach*. John Wiley & Sons, New York, NY.

7

Sampling and Estimation for Sustainability

Sampling approaches play an important role in providing information for making sustainability decisions under the Six Sigma methodology. Six Sigma is statistically based and requires good sampling strategies. This chapter presents a selection of basic sampling techniques.

Statistical Sampling

Sampling approaches provide essential information for making business decisions. Market researchers, for example, need to sample from a large base of customers or potential customers; auditors must sample among large numbers of transactions; and quality control analysts need to sample production output to verify quality levels. Most populations, even if they are finite, are generally too large to deal with effectively or practically. For instance, it would be impractical as well as too expensive to survey the entire population of consumers in the United States. Yet, we must draw inferences about the population. Thus, the purpose of sampling is to obtain sufficient information to draw a valid inference about a population.

Sampling Techniques

A *sample space* of an experiment is the set of all possible distinct outcomes of the experiment. An *experiment* is some process that generates distinct sets of observations. The simplest and most common example is the experiment of tossing a coin to observe whether heads or tails will show up. An *outcome* is a distinct observation resulting from a single trial of an experiment. In the experiment of tossing a coin, heads and tails are the two possible outcomes. Thus, the sample space consists of only two items.

There are several examples of statistical experiments suitable for project control. A simple experiment may involve checking to see whether it rains or not on a given day. Another experiment may involve counting how many tasks fall behind schedule during a project life cycle. An experiment may involve recording how long it takes to perform a given activity in each of several trials. The outcome of any experiment is frequently referred to as a *random outcome* because each outcome is independent and has the same chance of occurring. We cannot predict with certainty what the outcome of a particular trial of the experiment would be. An event can be a collection of outcomes.

Sample

A sample is a subset of a population that is selected for observation and statistical analysis. Inferences are drawn about the population based on the results of the analysis of the sample. The reasons for using sampling rather than complete population enumeration are as follows:

1. It is more economical to work with a sample.
2. There is a time advantage to using a sample.
3. Populations are typically too large to work with.
4. A sample is more accessible than the whole population.
5. In some cases, the sample may have to be destroyed during the analysis.

Sampling methods can be subjective or probabilistic. Subjective methods include judgment sampling, in which expert judgment is used to select the sample (survey the "best" customers), and convenient sampling, in which samples are selected based on the ease with which the data can be collected (survey all customers I happen to visit this month). Probabilistic sampling involves selecting the items in the sample using some random procedure. Probabilistic sampling is necessary to draw valid statistical conclusions. The most common probabilistic sampling approach is simple random sampling. Simple random sampling involves selecting items from a population so that every subset of a given size has an equal chance of being selected. If the population data are stored in a database, simple random samples can generally be obtained easily by generating random numbers.

Systematic Sampling

This is a sampling plan that selects items periodically from the population. For example, to sample 250 names from a list of 100,000, every 1600th name could be selected. If the first item is selected randomly among the first 1600, then the result will be a probabilistic sample. This approach can be used for telephone sampling when supported by an automatic dialer that is programmed to dial numbers in a systematic manner. However, systematic sampling is not the same as simple random sampling because for any sample, every possible sample of a given size in the population does not have an equal chance of being selected. In some situations this approach can induce significant bias if the population has some underlying pattern. For instance, sampling orders received every seven days may not yield a representative sample if customers tend to send orders on certain days every week. There are two options for sampling:

1. Sampling can be periodic (i.e., systematic), and you will be prompted for the period, which is the interval between sample observations from the beginning of the data set. For instance, if a period of 5 is used, observations 5, 10, 15, and so on, will be selected as samples.
2. Sampling can also be random. Tools and techniques are widely available for drawing random samples.

Stratified Sampling

This type of sampling applies to populations that are divided into natural subsets (strata) and allocates the appropriate proportion of samples to each stratum. For example, a large city may be divided into political districts called wards. Each ward has a different number of citizens. A stratified sample would choose a sample of individuals in each ward proportionate to its size. This approach ensures that each stratum is weighted by its size relative to the population and can provide better results than simple random sampling if the items in each stratum are not homogeneous. However, issues of cost or significance of certain strata might make a disproportionate sample more useful. For example, the ethnic or racial mix of each ward might be significantly different, making it difficult for a stratified sample to obtain the desired information.

Errors in Sampling

The purpose of sampling is to obtain statistics that estimate population parameters. Sample design can lead to two sources of errors. The first type of error, non-sampling error, occurs when the sample does not represent the target population adequately. This is generally a result of poor sample design, such as using a convenience sample when a simple random sample would have been more appropriate. Sampling (statistical) error occurs because samples are only a subset of the total population.

Sampling error is inherent in any sampling process, and although it can be minimized, it cannot be totally avoided. Sampling error depends on the size of the sample relative to the population. Thus, determining the number of samples to take is essentially a statistical issue that is based on the accuracy of the estimates needed to derive a useful conclusion. However, from a practical standpoint, one must also consider the cost of sampling and sometimes make a trade-off between cost and the information that is obtained.

Descriptive statistics. Descriptive statistics refers to analyses that are performed to describe the nature of a process or operation. The analyses presented in the previous section fall under the category of descriptive statistics because they are concerned with summary calculations and graphical displays of observations.

Inferential statistics. Inferential statistics refers to the process of drawing inferences about a process based on a limited observation of the process. The techniques presented in this section fall under the category of inferential statistics. Inferential statistics is of interest because it is dynamic and provides generalizations about a population by investigating only a portion of the population. The portion of the population investigated is referred to as a sample. As an example, the expected duration of a proposed task can be inferred from several previous observations of the durations of identical tasks.

Deductive statistics. Deductive statistics involves assigning properties to a specific item in a set based on the properties of a general class covering the set. For example, if it is known that 90% of projects in a given organization fail, then deduction can be used to assign a probability of 90% to the event that a specific project in the organization will fail.

Inductive statistics. Inductive statistics involves drawing general conclusions from specific facts. Inferences about populations are drawn from samples. For example, if 95% of a sample of 100 people surveyed in a 5,000-person organization favor a particular project, then induction can be used to conclude that 95% of the personnel in the organization favor the project. The different types of statistics play important roles in project control. Sampling is an important part of drawing inferences.

There are three primary types of samples. They differ in the manner in which their elementary units are chosen.

Convenience Sample. A convenience sample refers to a sample that is selected on the basis of how convenient certain elements of the population are for observation.

Judgment Sample. A judgment sample is one that is obtained based on the discretion of someone familiar with the relevant characteristics of the population.

Random Sample. A random sample refers to a sample whereby the elements of the sample are chosen at random. This is the most important type of sample for statistical analysis. In random sampling, all the items in the population have an equal chance of being selected for inclusion in the sample.

Since a sample is a collection of observations representing only a portion of the population, the way in which the sample is chosen can significantly affect the adequacy and reliability of the sample. Even after the sample is chosen, the manner in which specific observations are obtained may still affect the validity of the results. The possible bias and errors in the sampling process are discussed next.

Sampling Error

A sampling error refers to the difference between a sample mean and the population mean that is due solely to the particular sample elements that are selected for observation.

Nonsampling Error

A nonsampling error refers to an error that is due solely to the manner in which the observation is made.

Sampling Bias

A sampling bias refers to the tendency to favor the selection of certain sample elements having specific characteristics. For example, a sampling bias may occur if a sample of the personnel is selected from only the engineering department in a survey addressing the implementation of high technology projects.

Stratified Sampling

Stratified sampling involves dividing the population into classes, or groups, called strata. The items contained in each stratum are expected to be homogeneous with respect to the characteristics to be studied. A random subsample is taken from each stratum. The subsamples from all the strata are then combined to form the desired overall sample. Stratified sampling is typically used for a heterogeneous population such as data on employee productivity in an organization. Under stratification, groups of employees are selected so that the individuals within each stratum are mostly homogeneous and the strata are different from one another. As another example, a survey of project managers on some important issue of personnel management may be conducted by forming strata on the basis of the types of projects they manage. There may be one stratum for technical projects, one for construction projects, and one for manufacturing projects.

A *proportionate stratified sampling* results if the units in the sample are allocated among the strata in proportion to the relative number of units in each stratum in the population. That is, an equal sampling ratio is assigned to all strata in a proportionate stratified sampling. In *disproportionate stratified sampling,* the sampling ratio for each stratum is inversely related to the level of homogeneity of the units in the stratum. The more homogeneous the stratum, the smaller its proportion included in the overall sample. The rationale for using disproportionate stratified sampling is that when the units in a stratum are more homogeneous, a smaller subsample is needed to ensure good representation. The smaller subsample helps reduce sampling cost.

Cluster Sampling

Cluster sampling involves the selection of random clusters, or groups, from the population. The desired overall sample is made up of the units in each cluster. Cluster sampling is different from stratified sampling in that differences between clusters are usually small. In addition, the units within each cluster are generally more heterogeneous. Each cluster, also known as *primary sampling unit,* is expected to be a scaled-down model that gives a good representation of the characteristics of the population.

All the units in each cluster may be included in the overall sample or a subsample of the units in each cluster may be used. If all the units of the selected clusters are included in the overall sample, the procedure is referred to as *single-stage sampling.* If a subsample is taken at random from each selected cluster and all units of each subsample are included in the overall sample, then the sampling procedure is called *two-stage sampling.* If the sampling procedure involves more than two stages of subsampling, then the procedure is referred to as *multistage sampling.* Cluster sampling is typically less expensive to implement than stratified sampling. For example, the cost of taking a random sample of 2000 managers from different industry types may be reduced by first selecting a sample, or cluster, of 25 industries and then selecting 80 managers from each of the 25 industries. This represents a two-stage sampling that will be considerably cheaper than trying to survey 2,000 individuals in several industries in a single-stage procedure.

Once a sample has been drawn and observations of all the items in the sample are recorded, the task of data collection is completed. The next task involves organizing the raw data into a meaningful format. Frequency distribution is an effective tool for organizing data. Frequency distribution involves the arrangement of observations into classes so as to show the frequency of occurrences in each class.

Measurement Scales

Project control requires data collection, measurement, and analysis. In project management, the manager will encounter different types of measurement scales depending on the particular items to be controlled. Data may need to be collected on project schedules, costs, performance levels, problems, and so on. The different types of data measurement scales that are applicable are discussed next.

Nominal scale of measurement. A *nominal scale* is the lowest level of measurement scales. It classifies items into categories. The categories are mutually exclusive and collectively exhaustive. That is, the categories do not overlap and they cover all possible categories of the characteristics being observed. For example, in the analysis of the critical path in a project network, each job is classified as either critical or not critical. Gender, type of industry, job classification, and color are some examples of measurements on a nominal scale.

Ordinal scale of measurement. An *ordinal scale* is distinguished from a nominal scale by the property of order among the categories. An example is the process of prioritizing project tasks for resource allocation. We know that first is above second, but we do not know how far above. Similarly, we know that better is preferred to good, but we do not know by how much. In quality control, the ABC classification of items based on the Pareto distribution is an example of a measurement on an ordinal scale.

Interval scale of measurement. An *interval scale* is distinguished from an ordinal scale by having equal intervals between the units of measure. The assignment of priority ratings to project objectives on a scale of 0 to 10 is an example of a measurement on an interval scale. Even though an objective may have a priority rating of 0, it does not mean that the objective has absolutely no significance to the project team. Similarly, the scoring of 0 on an examination does not imply that a student knows absolutely nothing about the materials covered by the examination. Temperature is a good example of an item that is measured on an interval scale. Even though there is a zero point on the temperature scale, it is an arbitrary relative measure. Other examples of interval scales are IQ measurements and aptitude ratings.

Ratio scale of measurement. A *ratio scale* has the same properties of an interval scale but with a true zero point. For example, an estimate of a zero time unit for the duration of a task is a ratio scale measurement. Other examples of items measured on a ratio scale are cost, time, volume, length, height, weight, and inventory level. Many of the items measured in a project management environment will be on a ratio scale.

Another important aspect of data analysis for project control involves the classification scheme used. Most projects will have both *quantitative* and *qualitative* data. Quantitative data require that we describe the characteristics of the items being studied numerically. Qualitative data, on the other hand, are associated with object attributes that are not measured numerically. Most items measured on the nominal and ordinal scales will normally be classified into the qualitative data category while those measured on the interval and ratio scales will normally be classified into the quantitative data category.

The implication for project control is that qualitative data can lead to bias in the control mechanism because qualitative data are subject to the personal views and interpretations of the person using the data. Whenever possible, data for Six Sigma sustainability should be based on quantitative measurements.

There is a class of project data referred to as *transient data*. This is defined as a volatile set of data that is used for one-time decision making and is not then needed again. An example may be the number of operators that show up at a job site on a given day. Unless there is some correlation between the day-to-day attendance records of operators, this piece of information will have relevance only for that given day. The project manager can make his decision for that day on the basis of that day's attendance record. Transient data need not be stored in a permanent database unless it may be needed for future analysis or uses (e.g., forecasting, incentive programs, performance review).

Recurring data refers to data that is encountered frequently enough to necessitate storage on a permanent basis. An example is a file containing contract due dates. This file will need to be kept at least through the project life cycle. Recurring data may be further categorized into *static data* and *dynamic data*. A recurring data that is static will retain its original parameters and values each time it is retrieved and used. A recurring data that is dynamic has the potential for taking on different parameters and values each time it is retrieved and used. Storage and retrieval considerations for project control should address the following questions:

1. What is the origin of the data?
2. For how long will the data be maintained?
3. Who needs access to the data?
4. What will the data be used for?
5. How often will the data be needed?
6. Is the data for lookup purposes only (i.e., no printouts)?
7. Is the data for reporting purposes (i.e., generate reports)?
8. In what format is the data needed?
9. How fast will the data need to be retrieved?
10. What security measures are needed for the data?

Data Determination and Collection

It is essential to determine what data to collect for project control purposes. Data collection and analysis are basic components of generating information for project control. The requirements for data collection are discussed next.

Choosing the data. This involves selecting data on the basis of their relevance and the level of likelihood that they will be needed for future decisions and whether or not they contribute to making the decision better. The intended users of the data should also be identified.

Collecting the data. This identifies a suitable method of collecting the data as well as the source from which the data will be collected. The collection method will depend on the particular operation being addressed. The common methods include manual tabulation, direct keyboard entry, optical character reader, magnetic coding, electronic scanner, and more recently, voice command. An input control may be used to confirm the accuracy of collected data. Examples of items to control when collecting data are the following:

Relevance check. This checks if the data is relevant to the prevailing problem. For example, data collected on personnel productivity may not be relevant for a decision involving marketing strategies.

Limit check. This checks to ensure that the data is within known or acceptable limits. For example, an employee overtime claim amounting to over 80 hours per week for several weeks in a row is an indication of a record well beyond ordinary limits.

Critical value. This identifies a boundary point for data values. Values below or above a critical value fall in different data categories. For example, the lower specification limit for a given characteristic of a product is a critical value that determines whether the product meets quality requirements.

Coding the data. This refers to the technique used in representing data in a form useful for generating information. This should be done in a compact and yet meaningful format. The performance of information systems can be greatly improved if effective data formats and coding are designed into the system right from the beginning.

Processing the data. Data processing is the manipulation of data to generate useful information. Different types of information may be generated from a given data set depending on how it is processed. The processing method should consider how the information will be used, who will be using it, and what caliber of system response time is desired. If possible, processing controls should be used. This may involve the following:

Control total. Check for the completeness of the processing by comparing accumulated results to a known total. An example of this is the comparison of machine throughput to a standard production level or the comparison of cumulative project budget depletion to a cost accounting standard.

Consistency check. Check if the processing is producing the same results for similar data. For example, an electronic inspection device that suddenly shows a measurement that is 10 times higher than the norm warrants an investigation of both the input and the processing mechanisms.

Scales of measurement. For numeric scales, specify units of measurement, increments, the zero point on the measurement scale, and the range of values.

Using the information. Using information involves people. Computers can collect data, manipulate data, and generate information, but the ultimate decision rests with people, and decision making starts when information becomes available. Intuition, experience, training, interest, and ethics are just a few of the factors that determine how people use information. The same piece of information that is positively used to further the progress of a project in one instance may also be used negatively in another instance. To ensure that data and information are used appropriately, computer-based security measures can be built into the information system.

Project data may be obtained from several sources. Some potential sources are the following:

- Formal reports
- Interviews and surveys
- Regular project meetings
- Personnel time cards or work schedules

The timing of data is also very important for project control purposes. The contents, level of detail, and frequency of data can affect the control process. An important aspect of project management is the determination of the data required to generate the information needed for project control. The function of keeping track of the vast quantity of rapidly changing and interrelated data about project attributes can be very complicated. The major steps involved in data analysis for project control are as follows:

- Data collection
- Data analysis and presentation
- Decision making
- Implementation of action

Data is processed to generate information. Information is analyzed by the decision maker to make the required decisions. Good decisions are based on timely and relevant information, which in turn is based on reliable data. Data analysis for project control may involve the following functions:

- Organizing and printing computer-generated information in a form usable by managers
- Integrating different hardware and software systems to communicate in the same project environment
- Incorporating new technologies such as expert systems into data analysis
- Using graphics and other presentation techniques to convey project information

Proper data management will prevent misuse, misinterpretation, or mishandling. Data is needed at every stage in the life cycle of a project from the problem identification stage through the project phaseout stage. The various items for which data may be needed are project specifications, feasibility study, resource availability, staff size, schedule, project status, performance data, and phaseout plan. The documentation of data requirements should cover the following:

- Data summary. A data summary is a general summary of the information and decision for which the data is required as well as the form in which the data should be prepared. The summary indicates the impact of the data requirements on the organizational goals.
- Data processing environment. The processing environment identifies the project for which the data is required, the user personnel, and the computer system to be used in processing the data. It refers to the project request or authorization and relationship to other projects and specifies the expected data communication needs and mode of transmission.
- Data policies and procedures. Data handling policies and procedures describe policies governing data handling, storage, and modification and the specific procedures for implementing changes to the data. Additionally, they provide instructions for data collection and organization.
- Static data. A static data description describes that portion of the data that is used mainly for reference purposes and it is rarely updated.

- Dynamic data. A dynamic data description describes that portion of the data that is frequently updated based on the prevailing circumstances in the organization.
- Data frequency. The frequency of data update specifies the expected frequency of data change for the dynamic portion of the data, for example, quarterly. This data change frequency should be described in relation to the frequency of processing.
- Data constraints. Data constraints refer to the limitations on the data requirements. Constraints may be procedural (e.g., based on corporate policy), technical (e.g., based on computer limitations), or imposed (e.g., based on project goals).
- Data compatibility. Data compatibility analysis involves ensuring that data collected for project control needs will be compatible with future needs.
- Data contingency. A data contingency plan concerns data security measures in case of accidental or deliberate damage or sabotage affecting hardware, software, or personnel.

Point Estimates

The most common point estimates are the descriptive statistical measures of the data set of interest. The point estimates are used to estimate the population parameters.

Unbiased Estimators

It seems quite intuitive the sample mean should provide a good point estimate for the population mean. The sample variance is computed by the formula

$$s^2 = \sum_{i=1}^{n} \frac{(x_i - \overline{x})^2}{n-1}$$

whereas the population variance is computed by

$$\sigma^2 = \sum_{i=1}^{n} \frac{(x_i - \mu)^2}{N}$$

It is important that estimators truly estimate the population parameters they are supposed to estimate. Suppose that we perform an experiment in which we repeatedly sampled from a population and computed a point estimate for a population parameter. Each individual point estimate will vary from the population parameter; however, we would hope that the long-term average (expected value) of all possible point estimates would equal the population parameter. If the expected value of an estimator equals the population parameter it is intended to estimate, the estimator is said to be unbiased. If this is not true, the estimator is called biased and will not provide correct results.

Interval Estimates

An interval estimate provides a range within which we believe the true population parameter falls. For example, a Gallup poll might report that 56% of voters support a certain candidate with a margin of error of ±3%. We would conclude that the true percentage of voters that support the candidate is probably between 53% and 59%. Therefore, we would have a lot of confidence in predicting that the candidate would win a forthcoming election. If, however, the poll showed a 52% level of support with a margin of error of ±4%, we might not be as confident in predicting a win because the true percentage of supportive voters is probably somewhere between 48% and 56%.

A confidence interval (CI) is an interval estimate that also specifies the likelihood that the interval contains the true population parameter. This probability is called the level of confidence, denoted by $1 - \alpha$, and is usually expressed as a percentage. For example, we might state that "a 90% CJ for the mean is 10 ± 2." The value 10 is the point estimate calculated from the sample data, and 2 can be thought of as a margin for error. Thus, the interval estimate is [8, 12]. However, this interval may or may not include the true population mean. If we take a different sample, we will most likely have a different point estimate, say 10.4, which, given the same margin of error, would yield the interval estimate [8.4, 12.4]. Again, this may or may not include the true population mean. If we chose 100 samples, leading to 100 different interval estimates, we would expect that 90% of the level of confidence would contain the population mean. We would say we are "90% confident" that the interval we obtain from sample data contains the true population mean. Commonly used confidence levels are 90%, 95%, and 99%; the higher the confidence level, the more assurance we have that the interval contains the true population parameter. As the confidence level increases, the confidence interval becomes larger to provide higher levels of assurance. The standard normal distribution is measured in units of standard deviations. We will define z_α as the value from the standard normal distribution that provides an upper tail probability of α. That is, the area to the right of z_α is equal to α. Some common values that are used in practice include $z_{0.025} = 1.96$ and $z_{0.05} = 1.645$.

Data Analysis and Presentation

Data analysis refers to the various mathematical and graphical operations that can be performed on data to elicit the inherent information contained in the data. The manner in which project data is analyzed and presented can affect how the information is perceived by the decision maker. The examples presented in this section illustrate how basic data analysis techniques can be used to convey important information for project control.

In many cases, data is represented as the answer to direct questions such as the following: When is the project deadline? Who are the people assigned to the first task? How many resource units are available? Are enough funds available for the project? What are the quarterly expenditures on the project for the past two years? Is personnel productivity low, average, or high? Who is the person in charge of the project? Answers to these types of questions constitute data of different forms or expressed on different scales. The resulting data may be qualitative or quantitative. Different techniques are available for analyzing the different types of data. This section discusses some of the basic techniques for data analysis. The data presented in Table 7.1 is used to illustrate the data analysis techniques.

Raw Data

Raw data consists of ordinary observations recorded for a decision variable or factor. Examples of factors for which data may be collected for decision making are revenue, cost, personnel productivity, task duration, project completion time, product quality, and resource availability. Raw data should be organized into a format suitable for visual review and computational analysis. The data in Table 7.1 represents the quarterly revenues from projects A, B, C, and D. For example, the data for quarter 1 indicates that project C yielded the highest revenue of $4,500,000 while project B yielded the lowest revenue of $1,200,000. Figure 7.1 presents the raw data of project revenue as a line graph. The same information is presented as a multiple bar chart in Figure 7.2.

TABLE 7.1

Quarterly Revenue from Four Projects (in $1,000s)

Project	Quarter 1	Quarter 2	Quarter 3	Quarter 4	Row Total
A	3,000	3,200	3,400	2,800	12,400
B	1,200	1,900	2,500	2,400	8,000
C	4,500	3,400	4,600	4,200	16,700
D	2,000	2,500	3,200	2,600	10,300
Column total	10,700	11,000	13,700	12,000	47,400

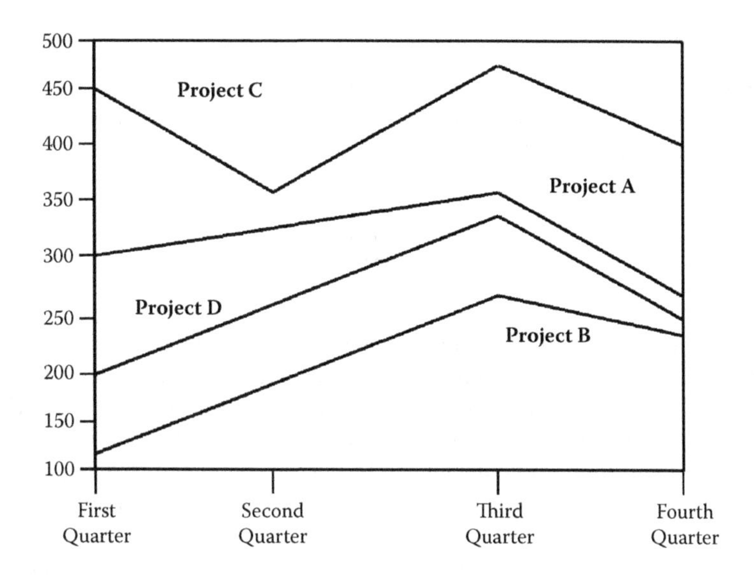

FIGURE 7.1
Line graph of quarterly project revenues.

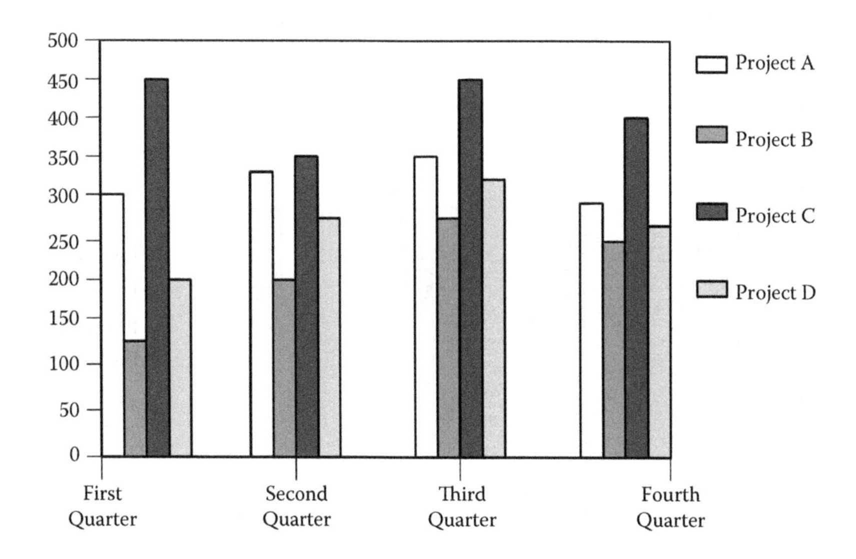

FIGURE 7.2
Multiple bar chart of quarterly project revenues.

Total Revenue

A total or sum is a measure that indicates the overall effect of a particular variable. If $X_1, X_2, X_3, \ldots, X_n$ represent a set of n observations (e.g., revenues), then the total is computed as

$$T = \sum_{i=1}^{n} X_i$$

For the data in Table 7.1, the total revenue for each project is shown in the last column. The totals indicate that project C brought in the largest total revenue over the four quarters under consideration while project B produced the lowest total revenue. The last row of the table shows the total revenue for each quarter. The totals reveal that the largest revenue occurred in the third quarter. The first quarter brought in the lowest total revenue. The grand total revenue for the four projects over the four quarters is shown as $47,400,000 in the last cell in the table. Figure 7.3 presents the quarterly total revenues as stacked bar charts. Each segment in a stack of bars represents the revenue contribution from a particular project. The total revenues for the four projects over the four quarters are shown in a pie chart in Figure 7.4. The percentage of the overall revenue contributed by each project is also shown on the pie chart.

Average Revenue

Average is one of the most used measures in data analysis. Given n observations (e.g., revenues), $X_1, X_2, X_3, \ldots, X_n$, the average of the observations is computed as

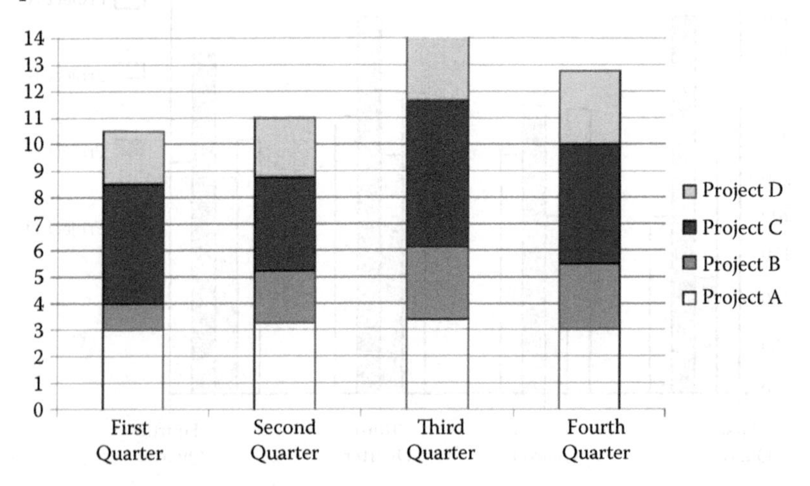

FIGURE 7.3
Stacked bar graph of quarterly total revenues.

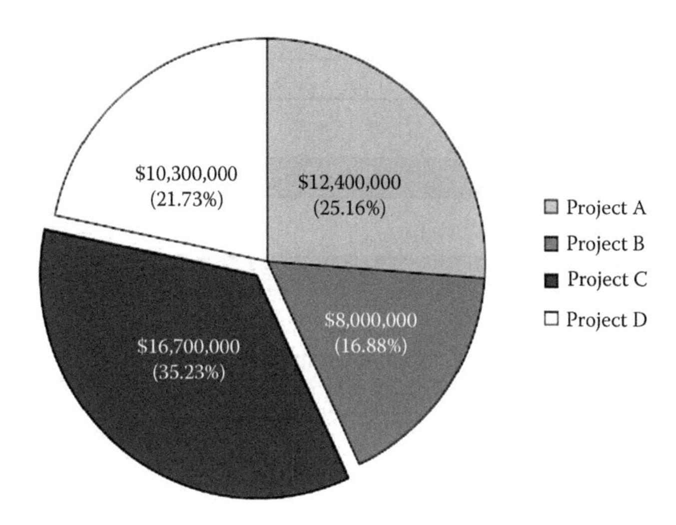

FIGURE 7.4
Pie chart of total revenue per project.

$$\bar{X} = \frac{\sum_{i=1}^{n} X_i}{n}$$

$$= \frac{T_x}{n}$$

where T_x is the sum of n revenues. For our sample data, the average quarterly revenues for the four projects are computed below:

$$\bar{X}_A = \frac{(3,000+3,200+3,400+2,800)(\$1,000)}{4} = \$3,100,000.00$$

$$\bar{X}_B = \frac{(1,200+1,900+2,500+2,400)(\$1,000)}{4} = \$2,000,000.00$$

$$\bar{X}_C = \frac{(4,500+3,400+4,600+4,200)(\$1,000)}{4} = \$4,175,000.000$$

$$\bar{X}_D = \frac{(2,000+2,500+3,200+2,600)(\$1,000)}{4} = \$2,575,000.00$$

Similarly, the expected average revenues per project for the four quarters are computed below:

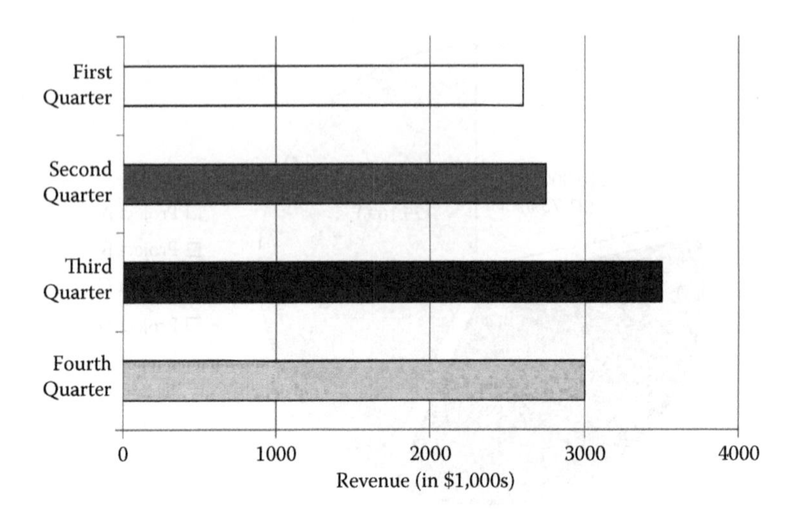

FIGURE 7.5
Average revenue per project for each quarter.

$$\bar{X}_1 = \frac{(3,000 + 1,200 + 4,500 + 2,000)(\$1,000)}{4} = \$2,675,000.00$$

$$\bar{X}_2 = \frac{(3,200 + 1,900 + 3,400 + 2,500)(\$1,000)}{4} = \$2,750,000.00$$

$$\bar{X}_3 = \frac{(3,400 + 2,500 + 4,600 + 3,200)(\$1,000)}{4} = \$3,425,000.00$$

$$\bar{X}_4 = \frac{(2,800 + 2,400 + 4,200 + 2,600)(\$1,000)}{4} = \$3,000,000.00$$

The above values are shown in a bar chart in Figure 7.5. The average revenue from any of the four projects in any given quarter is calculated as the sum of all the observations divided by the number of observations. That is,

$$\bar{\bar{X}} = \frac{\sum_{i=1}^{N} \sum_{j=1}^{M} X_{ij}}{K}$$

where

N = number of projects
M = number of quarters
K = total number of observations ($K = NM$)

The overall average per project per quarter is computed as follows:

$$\bar{\bar{X}} = \frac{\$47,400,000}{16} = \$47,400.00$$

As a cross-check, the sum of the quarterly averages should be equal to the sum of the project revenue averages, which is equal to the grand total divided by 4.

$$(2{,}675 + 2{,}750 + 3{,}425 + 3{,}000)(\$1{,}000) = (3{,}100 + 2{,}000 + 4{,}175 + 2{,}575)(\$1{,}000)$$

$$= \$11{,}800.000 = \$47{,}400{,}000/4$$

The cross-check procedure above works because we have a balanced table of observations. That is, we have four projects and four quarters. If there were only three projects, for example, the sum of the quarterly averages would not be equal to the sum of the project averages.

Median Revenue

The median is the value that falls in the middle of a group of observations arranged in order of magnitude. One-half of the observations are above the median and the other half are below the median. The method of determining the median depends on whether the observations are organized into a frequency distribution. For unorganized data, it is necessary to arrange the data in an increasing or decreasing order before finding the median. Given K observations (e.g., revenues), X_1, X_2, X_3, …, X_K, arranged in increasing or decreasing order, the median is identified as the value in position $(K + 1)/2$ in the data arrangement if K is an odd number. If K is an even number, then the average of the two middle values is considered to be the median. If the sample data are arranged in increasing order, we would get the following:

1,200, 1,900, 2,000, 2,400, 2,500, 2,500, 2,600, 2,800, 3,000, 3,200, 3,200, 3,400, 3,400, 4,200, 4,500, 4,600

The median is then calculated as (2,800 + 3,000)/2 = 2,900. Half of the recorded revenues are expected to be above $2,900,000 while half are expected to be below that amount. Figure 7.6 presents a bar chart of the revenue data arranged in increasing order. The median is anywhere between the eighth and ninth values in the ordered data.

Quartiles and Percentiles

The median is a position measure because its value is based on its position in a set of observations. Other measures of position are *quartiles* and *percentiles*.

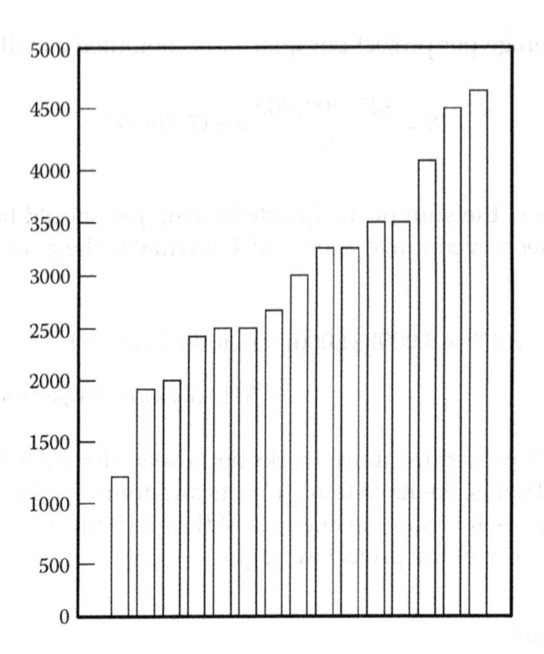

FIGURE 7.6
Bar chart of ordered data.

There are three quartiles, which divide a set of data into four equal categories. The first quartile, denoted Q_1, is the value below which one-fourth of all the observations in the data set fall. The second quartile, denoted, Q_2, is the value below which two-fourths or one-half of all the observations in the data set fall. The third quartile, denoted Q_3, is the value below which three-fourths of the observations fall. The second quartile is identical to the median. It is technically incorrect to talk of the fourth quartile because that will imply that there is a point within the data set below which all the data points fall: a contradiction! A data point cannot lie within the range of the observations and at the same time exceed all the observations, including itself.

The concept of percentiles is similar to the concept of quartiles except that reference is made to percentage points. There are 99 percentiles that divide a set of observations into 100 equal parts. The *X* percentile is the value below which *X* percent of the data fall. The 99 percentile refers to the point below which 99% of the observations fall. The three quartiles discussed previously are regarded as the 25th, 50th, and 75th percentiles. It would be technically incorrect to talk of the 100th percentile. In performance ratings, such as on an examination or product quality level, the higher the percentile of an individual or product, the better. In many cases, recorded data are classified into categories that are not indexed to numerical measures. In such cases, other measures of central tendency or position will be needed. An example of such a measure is the mode.

The Mode

The mode is defined as the value that has the highest frequency in a set of observations. When the recorded observations can be classified only into categories, the mode can be particularly helpful in describing the data. Given a set of K observations (e.g., revenues), $X_1, X_2, X_3, \ldots, X_K$, the mode is identified as that value that occurs more than any other value in the set. Sometimes, the mode is not unique in a set of observations. For example, in Table 7.1, $2,500, $3,200, and $3,400 all have the same number of occurrences. Each of them is a mode of the set of revenue observations. If there is a unique mode in a set of observations, then the data is said to be unimodal. The mode is very useful in expressing the central tendency for observations with qualitative characteristics such as color, marital status, or state of origin. The three modes in the raw data can be identified in Figure 6.1.

Range of Revenue

The range is determined by the two extreme values in a set of observations. Given K observations (e.g., revenues), $X_1, X_2, X_3, \ldots, X_K$, the range of the observations is simply the difference between the lowest and the highest observations. This measure is useful when the analyst wants to know the extent of extreme variations in a parameter. The range of the revenues in our sample data is ($4,600,000 – $1,200,000) = $3,400,000. Because of its dependence on only two values, the range tends to increase as the sample size increases. Furthermore, it does not provide a measurement of the variability of the observations relative to the center of the distribution. This is why the standard deviation is normally used as a more reliable measure of dispersion than the range.

The variability of a distribution is generally expressed in terms of the deviation of each observed value from the sample average. If the deviations are small, the set of data is said to have low variability. The deviations provide information about the degree of dispersion in a set of observations. A general formula to evaluate the variability of data cannot be based on the deviations. This is because some of the deviations are negative while some are positive and the sum of all the deviations is equal to zero. One possible solution to this is to compute the average deviation.

Average Deviation

The average deviation is the average of the absolute values of the deviations from the sample average. Given K observations (e.g., revenues), $X_1, X_2, X_3, \ldots, X_K$, the average deviation of the data is computed as follows:

$$\bar{D} = \frac{\sum_{i=1}^{K} |X_i - \bar{X}|}{K}$$

TABLE 7.2

Computation of Average Deviation, Standard Deviation, and Variance

Observation Number (i)	Recorded Observation X_i	Deviation from Average $X_i - \bar{X}$	Absolute Value $\lvert X_i - \bar{X} \rvert$	Square of Deviation $(X_i - \bar{X})^2$
1	3,000	37.5	37.5	1,406.25
2	1,200	–1762.5	1762.5	3,106,406.30
3	4,500	1537.5	1537.5	2,363,906.30
4	2,000	–962.5	962.5	926,406.25
5	3,200	237.5	237.5	56,406.25
6	1,900	–1062.5	1062.5	1,128,906.30
7	3,400	437.5	437.5	191,406.25
8	2,500	–462.5	462.5	213,906.25
9	3,400	437.5	437.5	191,406.25
10	2,500	–462.5	462.5	213,906.25
11	4,600	1637.5	1637.5	2,681,406.30
12	3,200	237.5	237.5	56,406.25
13	2,800	–162.5	162.5	26,406.25
14	2,400	–562.5	562.5	316,406.25
15	4,200	1237.5	1237.5	1,531,406.30
16	2,600	–362.5	362.5	131,406.25
Total	47,400.0	0.0	11,600.0	13,137,500.25
Average	2,962.5	0.0	725.0	821,093.77
Square root				906.14

Table 7.2 shows how the average deviation is computed for our sample data. One aspect of the average deviation measure is that the procedure ignores the sign associated with each deviation. Despite this disadvantage, its simplicity and ease of computation make it useful. In addition, knowledge of the average deviation helps in understanding the standard deviation, which is the most important measure of dispersion available.

Sample Variance

Sample variance is the average of the squared deviations computed from a set of observations. If the variance of a set of observations is large, the data is said to have a large variability. For example, a large variability in the levels of productivity of a project team may indicate a lack of consistency or improper methods in the project functions. Given K observations (e.g., revenues), $X_1, X_2, X_3, \ldots, X_K$, the sample variance of the data is computed as

$$s^2 = \frac{\sum_{i=1}^{K} \left(X_i - \bar{X}\right)^2}{K-1}$$

The variance can also be computed by the following alternate formulas:

$$s^2 = \frac{\sum_{i=1}^{K}\left(X_i^2 - \left(\frac{1}{K}\right)\right)\left[\sum_{i=1}^{K} X_i\right]^2}{K-1}$$

$$s^2 = \frac{\sum_{i=1}^{K} X_i^2 - K\left(\bar{X}^2\right)}{K-1}$$

Using the first formula, the sample variance of the data in Table 7.2 is calculated as

$$s^2 = \frac{13{,}137{,}500.25}{16-1} = 875{,}833.33$$

The average calculated in the last column of Table 7.2 is obtained by dividing the total for that column by 16 instead of 16 – 1 = 15. That average is not the correct value of the sample variance. However, as the number of observations gets very large, the average as computed in the table will become a close estimate for the correct sample variance. Analysts make a distinction between the two values by referring to the average calculated in the table as the population variance when K is very large and referring to the average calculated by the formulas above as the sample variance particularly when K is small. For our example, the population variance is given by

$$\sigma^2 = \frac{\sum_{i=1}^{K}\left(X_i - \bar{X}\right)^2}{K}$$

$$= \frac{13{,}137{,}500.25}{16} = 821{,}093.77$$

while the sample variance, as shown previously for the same data set, is given by

$$\sigma^2 = \frac{\sum_{i=1}^{K}\left(X_i - \bar{X}\right)^2}{K-1}$$

$$= \frac{13{,}137{,}500.25}{(16-1)} = 875{,}833.33$$

Standard Deviation

The sample standard deviation of a set of observations is the positive square root of the sample variance. The use of variance as a measure of variability has some drawbacks. For example, the knowledge of the variance is helpful only when two or more sets of observations are compared. Because of the squaring operation, the variance is expressed in square units rather than the original units of the raw data. To get a reliable feel for the variability in the data, it is necessary to restore the original units by performing the square root operation on the variance. This is why standard deviation is a widely recognized measure of variability. Given K observations (e.g., revenues), $X_1, X_2, X_3, \ldots, X_K$, the sample standard deviation of the data is computed as

$$s = \sqrt{\frac{\sum_{i=1}^{K}(X_i - \bar{X})^2}{K-1}}$$

As in the case of the sample variance, the sample standard deviation can also be computed by the following alternate formulas:

$$s = \sqrt{\frac{\sum_{i=1}^{K} X_i^2 - \left(\frac{1}{K}\right)\left[\sum_{i=1}^{K} X_i\right]^2}{K-1}}$$

$$s = \sqrt{\frac{\sum_{i=1}^{K} X_i^2 - K(\bar{X})^2}{K-1}}$$

Using the first formula, the sample standard deviation of the data in Table 7.2 is calculated as follows:

$$s = \sqrt{\frac{13,137,500.25}{(16-1)}} = \sqrt{875,833.33} = 935.8597$$

We can say that the variability in the expected revenue per project per quarter is $935,859.70. The population sample standard deviation is given by

$$\sigma = \sqrt{\frac{\sum_{i=1}^{K}(X_i - \bar{X})^2}{K}}$$

$$= \sqrt{\frac{13,137,500.25}{16}} = 906.1423$$

while the sample standard deviation is given by

$$s = \sqrt{\frac{\sum_{i=1}^{K}(X_i - \bar{X})^2}{K-1}}$$

$$= \sqrt{\frac{13,137,500.25}{(16-1)}} = 935.8597$$

The results of data analysis can be reviewed directly to determine where and when project control actions may be needed. The results can also be used to generate control charts.

Calculation Example

Suppose a set of data is collected about project costs in an organization. Twenty projects are selected for the study. The observations below are recorded in thousands of dollars:

$3,000	$1,100	$4,200	$800	$3,000
$1,800	$2,500	$2,500	$1,700	$3,000
$2,900	$2,100	$2,300	$2,500	$1,500
$3,500	$2,600	$1,300	$2,100	$3,600

Table 7.3 shows the tabulation of the cost data as a frequency distribution. Note how the end points of the class intervals are selected such that no recorded data point falls at an end point of a class. Note also that seven class intervals seem to be the most appropriate size for this particular

TABLE 7.3

Frequency Distribution of Project Cost Data

Cost Interval ($)	Midpoint ($)	Frequency	Cumulative Frequency
750–1,250	1,000	2	2
1,250–1,750	1,500	3	5
1,750–2,250	2,000	3	8
2,250–2,750	2,500	5	13
2,750–3,250	3,000	4	17
3,250–3,750	3,500	2	19
3,750–4,250	4,000	1	20
Total		20	

TABLE 7.4

Relative Frequency Distribution of Project Cost Data

Cost Interval ($)	Midpoint ($)	Frequency	Cumulative Frequency
750–1,250	1,000	0.10	0.10
1,250–1,750	1,500	0.15	0.25
1,750–2,250	2,000	0.15	0.40
2,250–2,750	2,500	0.25	0.65
2,750–3,250	3,000	0.20	0.85
3,250–3,750	3,500	0.10	0.95
3,750–4,250	4,000	0.05	1.00
Total		1.00	

set of observations. Each class internal has a spread of $500 which is an approximation obtained from the expression presented below.

$$W = \frac{X_{max} - X_{min}}{N}$$

$$= \frac{4200 - 800}{7} = 485.71 \approx 500$$

Table 7.4 shows the relative frequency distribution. The relative frequency of any class is the proportion of the total observations that fall into that class. It is obtained by dividing the frequency of the class by the total number of observations. The relative frequency of all the classes should add up to 1. From the relative frequency table, it is seen that 25% of the observed project costs fall within the range of $2,250 and $2,750. It is also noted that only 15% (0.10 + 0.05) of the observed project costs fall in the upper two intervals of project costs.

Figure 7.7 shows the histogram of the frequency distribution for the project cost data. Figure 7.8 presents a plot of the relative frequency of the project cost data. The plot of the cumulative relative frequency is superimposed on the relative frequency plot.

The relative frequency of the observations in each class represents the probability that a project cost will fall within that range of costs. The corresponding cumulative relative frequency gives the probability that project cost will fall below the midpoint of that class interval. For example, 85% of project costs in this example are expected to fall below or equal $3,000.

Diagnostic Tools

To facilitate data analysis for diagnosing a project for control purposes, we recommend using available graphical tools. The tools include flowcharts,

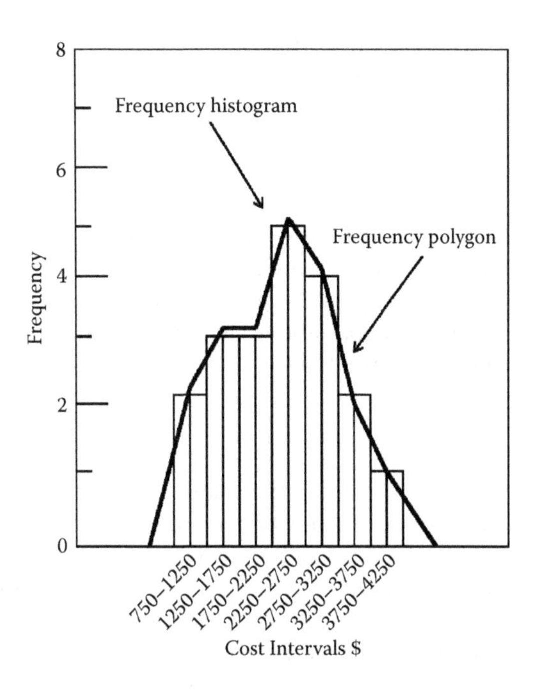

FIGURE 7.7
Histogram of project cost distribution data.

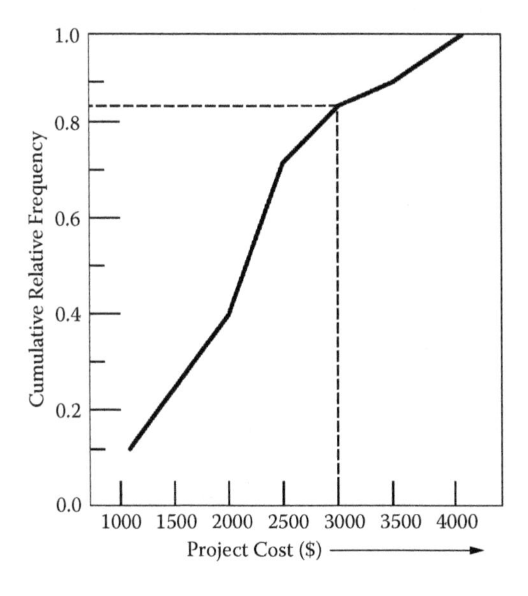

FIGURE 7.8
Plot of cumulative relative frequency.

Pareto diagrams, cause-and-effect diagrams, check sheets, scatter plots, run charts, and histograms. The tools are very effective for identifying problems that may need control actions.

Flowcharts

A flowchart is used to show the steps that a product or service follows from the beginning to the end of the process. It helps locate the value-added parts of the process steps. It also helps in locating the unnecessary steps in the process where unnecessary cost and labor exist. These unnecessary steps can be reduced or permanently eliminated.

Pareto Diagram

A Pareto diagram is used to display the relative importance or size of problems to determine the order of priority for projects. It can help identify the projects to concentrate on. For example, in Figure 7.9, analysts may tend to focus on project I since this is where the greatest dollar loss occurs. The criticality of a project may be determined by a combination of factors. The selection of project I as the most critical project to focus on should not be made solely on the largest dollar loss alone or any other single criterion. For example, if project I involves determining the number of accidents per year and project II involves determining the number of deaths per year, then project II may have priority since focusing on the number of deaths may be more critical than focusing on the number of accidents, even though the frequency of accidents is more than the frequency of deaths.

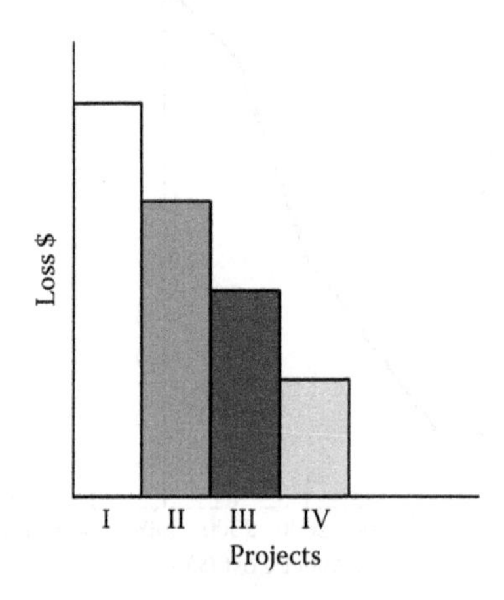

FIGURE 7.9
Relative dollar losses of quality improvement projects.

Scatter Plots

A scatter plot is used to study the relationships between two variables. It is sometimes called an X-Y plot. The plot gives a visual assessment of the location tendencies of data points. The appearance of a scatter plot can help identify the type of statistical analyses that may be needed for the data. For example, in regression analysis, a scatter plot can help an analyst determine the type of models to be investigated.

Run Charts and Check Sheets

A run chart is a tool that can be used to monitor the trends in a process over time. A check sheet is a preprinted table layout that facilitates data collection. Items to be recorded are preprinted in the table. Observations are recorded by simply checking appropriate cells in the table. A check sheet helps to automatically organize data for subsequent analysis. If properly designed, a check sheet can eliminate the need for counting data points during data analysis.

Histogram

A histogram is used to display the distribution of data by organizing the data points into evenly spaced numerical groupings that show the frequency of values in each group. Histograms can be used for quickly assessing the variation and distribution affecting a project. Important guidelines for drawing histograms are as follows:

Step 1: Determine the minimum and maximum values to be covered by the histogram.

Step 2: Select a number of histogram classes between 6 and 15. Having too few or too many classes will make it impossible to identify the underlying distribution.

Step 3: Set the same interval length for the histogram classes such that every observation in the data set falls within some class. The difference between midpoints of adjacent classes should be constant and equal to the length of each interval. If N represents the number of histogram classes, determine the interval length as shown below:

$$W = \frac{X_{max} - X_{min}}{N}$$

where X_i represents an observation in the data set.

Step 4: Count the number of observations that fall within each histogram class. This can be done by using a check sheet or any other counting technique.

Step 5: Draw a bar for each histogram class such that the height of the bar represents the number of observations in the class. If desired, the heights can be converted to relative proportions in which the height of each bar represents the percentage of the data set that falls within the histogram class.

The number of classes should not be so small or so large that the true nature of the underlying distribution cannot be identified. Generally, the number of classes should be between 6 and 20. The interval length of each class should be the same. The interval length should be selected such that every observation falls within some class. The difference between midpoints of adjacent classes should be constant and equal to the length of each interval.

A frequency polygon may be obtained by drawing a line to connect the midpoints at the top of the histogram bars. The polygon will show the spread and shape of the distribution of the data set. Three possible patterns of distribution may be revealed by the polygon: *symmetrical, positively skewed,* and *negatively skewed.* In a symmetrical distribution, the two halves of the graph are identical. In a positively skewed distribution (skewed to the right), there is a long tail stretching to the right side of the distribution. In a negatively skewed distribution (skewed to the left), there is a long tail stretching to the left side of the distribution.

Calculations under the Normal Curve

Most of the analyses involving the normal curve are done in the standardized domain. This is done by using the following transformation expression:

$$Z = \frac{X - \mu}{\sigma}$$

where Z is the standard normal random variable and X is the general normal random variable with a mean of μ and standard deviation of σ. The variable Z is often referred to as the *normal deviate.* One important aspect of the normal distribution relates to the percent of observations within one, two, or three standard deviations. Approximately 68.27% of observations following a normal distribution lie within plus and minus one standard deviation from the mean. Approximately 95.45% of the observations lie within plus or minus two standard deviations from the mean, and approximately 99.73% of the observations lie within plus or minus three standard deviations from the mean. These are shown graphically in Figure 7.10.

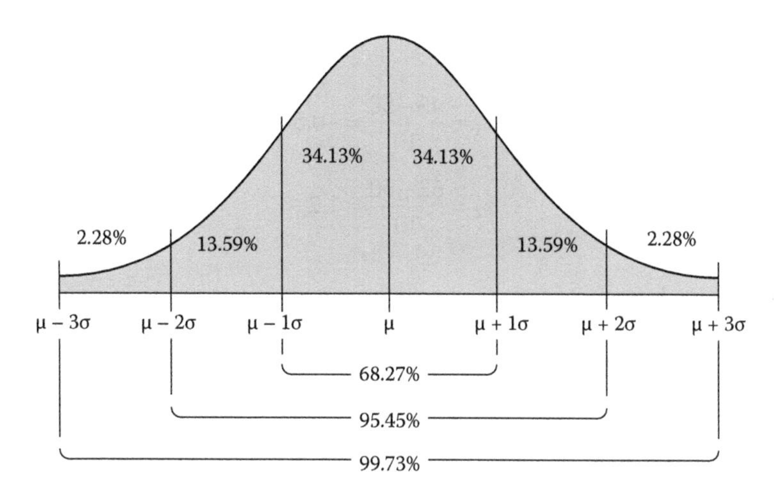

FIGURE 7.10
Areas under the normal curve.

To obtain probabilities for particular values of a random variable, it is necessary to know the probability distribution of the random variable. Because of the infinite possible combinations of means and standard deviation values, there are an infinite number of normal distributions. It is quite impractical to try and calculate probabilities directly from each one of them individually. The standard normal distribution can be applied to each and every possible normal random variable by using the transformation expression presented earlier. The standard normal distribution is of great importance in practice because it can be used to approximate many of the other discrete and continuous random variables.

Because the normal distribution represents a continuous random variable, it is impossible to calculate the probability of a single point on the curve. To determine probabilities, it is necessary to refer to intervals, such as the interval between point a and point b. The area under the curve from a to b represents the probability that the random variable will lie between a and b. That probability is calculated as follows:

Given:

Normal random variable X, representing task duration

Mean of $X = 50$ days

Standard deviation of $X = 10$

Required: The probability that the task duration will lie between 45 and 62 days

Solution: Let $X_1 = 45$ and $X_2 = 62$

Then,

$$z_1 = \frac{45-50}{10} = -0.5$$

$$z_2 = \frac{62-50}{10} = 1.2$$

Therefore, we have

$$P\,(45 < X < 62) = P\,(-0.5 < Z < 1.2)$$

$$= P\,(Z < 1.2) - P\,(Z < -0.5)$$

$$= P\,(Z < 1.2) - [1 - P\,(Z < 0.5)$$

$$= 0.8849 - (1 - 0.6915) = 0.5764$$

There is a 57.64% chance that this particular task will last between 45 and 62 days. Note that the area under the curve between 45 and 62 is calculated by first finding the total area to the left of 62 (i.e., 0.8849) and then subtracting the area to the left of 45 (i.e., 0.3085). Note also that $P\,(Z < -0.5)$ can be computed as $1 - P\,(Z < 0.5)$ for the case where the normal table does not contain negative values of z. The respective probabilities are read off the normal probability table given in Appendix A. To illustrate the use of the table, let us find the probability that Z will be less than 1.23. First, we locate the value of z equal to 1.2 in the left column of the table and then move across the row to the column under 0.03, where we read the value of 0.8907 inside the body of the table. Thus, $P\,(Z < 1.23) = 0.8907$. Using a *similar* process, the following additional examples are presented:

$$P(X < 55) = P\left(Z < \frac{55-50}{10}\right) = P(Z < 0.5) = 0.6915$$

$$P(X > 65) = P\left(Z > \frac{65-50}{10}\right)$$

$$= P(Z > 1.5) = 1 - P(Z < 1.5) = 1 - 0.9332 = 0.0668$$

Note that since the normal distribution table is constructed as cumulative probabilities from the left, $P\,(Z > 65)$ is calculated as $1 - P\,(Z < 65)$. Note that

$$P\ (Z < k) = 1.0, \text{ for any value } k \text{ that is greater than } 3.5$$

$$P\ (Z < 0) = 0.5$$

$$P\ (Z < k) = 0.0, \text{ for any value } k \text{ that is less than } -3.5$$

Probabilistic and statistical analyses offer a robust approach to evaluating project performance. Measurement, evaluation, and control actions may be influenced by probabilistic events. For example, resource allocation decision problems under uncertainty can be handled by appropriate decision tree models. With the statistical approach, the overall function of project control can be improved.

8

Managing Sustainability Projects

Sustainability projects must be managed the same way that any conventional projects are managed. The existing tools and techniques of project management as presented by Badiru (2012) are directly applicable to sustainability projects. Sustainability requires communication, cooperation, and coordination as a foundation for commitment and project discipline. Sustainability is achieved through management actions of humans. Human behavioral change is a prerequisite for achieving sustainability within the global realm of operations. Abraham Lincoln was the first to say, "The best way to predict your future is to create it."

This means that we have to take proactive actions in the pursuit of sustainability. We cannot afford to wait for the future to unfolded so that we can react to it. By using project management tools and techniques, we can plan, organize, schedule, and control sustainability projects. This chapter presents the fundamentals of project management. The tools and techniques can be adapted to fit specific sustainability projects, whether it is at home or at work. Project management is the pursuit of organizational goals within the constraints of time, cost, and quality expectations. This can be summarized into a few basic questions, such as the following:

- What needs to be done?
- What can be done?
- What will be done?
- Who will do it?
- When will it be done?
- Where will it be done?
- How will it be done?

These three factors of time, cost, and quality are synchronized to answer the above questions. The factors must be managed and controlled within the constraints of the iron triangle depicted in Figure 8.1. In this case, quality represents the composite collection of project requirements. In a situation where precise optimization is not possible, there will have to be trade-offs between these three factors of success. The concept of iron triangle is that a rigid triangle of constraints encases the project. Everything must be accomplished within the boundaries of time, cost, and quality. If better quality is

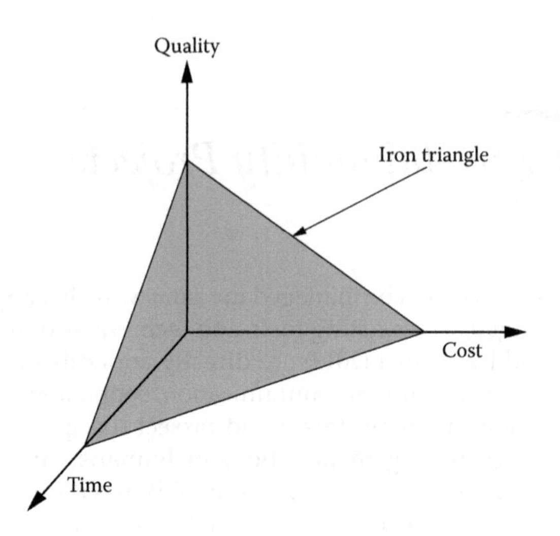

FIGURE 8.1
Project constraints of cost, time, and quality.

expected, a compromise along the axes of time and cost must be executed, thereby altering the shape of the triangle.

The trade-off relationships are not linear and must be visualized in a multidimensional context. This is better articulated by a 3-D view of the systems constraints as shown in Figure 8.2. Scope requirements determine the project boundary and trade-offs must be done within that boundary. If we label the eight corners of the box as (a), (b), (c), …, (h), we can iteratively assess the best operating point for the project. For example, we can address the following two operational questions:

1. From the point of view of the project sponsor, which corner is the most desired operating point in terms of combination of requirements, time, and cost?
2. From the point of view of the project executor, which corner is the most desired operating point in terms of combination of requirements, time, and cost?

Note that all the corners represent extreme operating points. We notice that point (e) is the do-nothing state, where there are no requirements, no time allocation, and no cost incurrence. This cannot be the desired operating state of any organization that seeks to remain productive. Point (a) represents an extreme case of meeting all requirements with no investment of time or cost allocation. This is an unrealistic extreme in any practical environment. It represents a case of getting something for nothing. Yet, it is the most desired operating point for the project sponsor. By comparison, point (c) provides

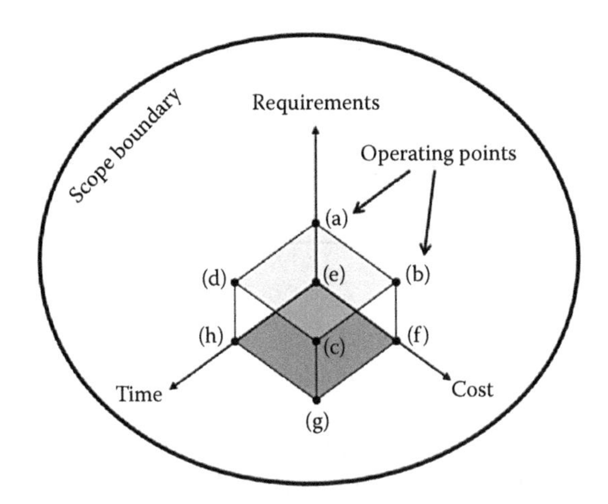

FIGURE 8.2
Systems constraints of cost, time, and quality within iron triangle.

the maximum possible for requirements, cost, and time. In other words, the highest levels of requirements can be met if the maximum possible time is allowed and the highest possible budget is allocated. This is an unrealistic expectation in any resource-conscious organization. You cannot get everything you ask for to execute a project. Yet, it is the most desired operating point for the project executor. Considering the two extreme points of (a) and (c), it is obvious that the project must be executed within some compromise region within the scope boundary. Figure 8.3 shows a possible view of a compromise surface with peaks and valleys representing give-and-take trade-off points within the constrained box. With proper control strategies, the project team can guide the project in the appropriate directions. The challenge is to come up with some analytical modeling technique to guide decision making

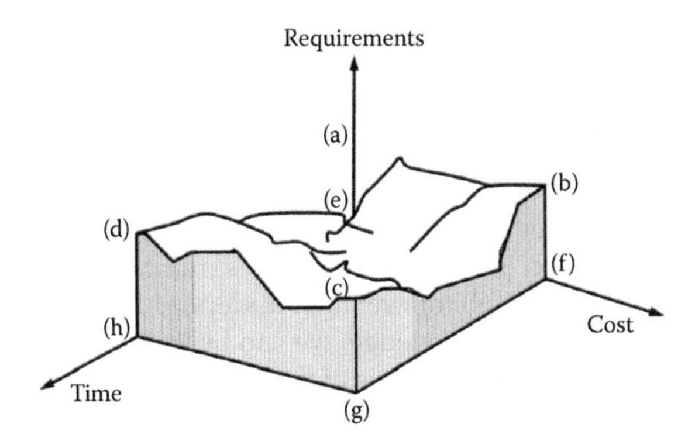

FIGURE 8.3
Compromise surface for cost, time, and requirements trade-off.

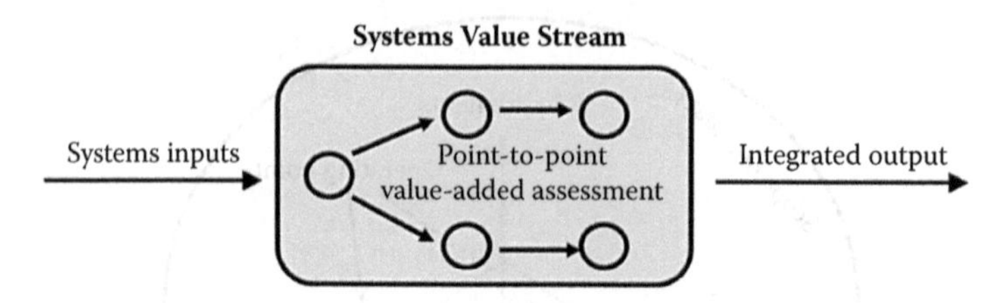

FIGURE 8.4
Systems value-stream structure.

over the compromise region. If we could collect sets of data over several repetitions of identical projects, then we could model a decision surface that can guide future executions of similar projects. Such typical repetitions of an identical project are most readily apparent in construction projects—for example, residential home development projects.

Organizations often inadvertently fall into an unstructured management "BLOBS" because it is simple, low-cost, and less time-consuming, until a problem develops. A desired alternative is to model the project system using a systems value-stream structure as shown in Figure 8.4. This uses a proactive and problem-preempting approach to execute projects. This alternative has the following advantages:

- Problem diagnosis is easier.
- Accountability is higher.
- Operating waste is minimized.
- Conflict resolution is faster.
- Value points are traceable.

Why Projects Fail

In spite of concerted efforts to maintain control of a project, many projects still fail. If no project ever fails, it would be a perfect world. But we all know that there is no perfection in human existence. To maintain a better control of a project, albeit no perfect control, we must understand the common reasons that projects fail. With this knowledge, we can better preempt project problems. Below are some common causes of project failure:

- Lack of communication
- Lack of cooperation

- Lack of coordination
- Diminished interest
- Diminished resources
- Change of objectives
    - Change of stakeholders
    - Change of priority
    - Change of perspectives
- Change of ownership
    - Change of scope
    - Budget cut
    - Shift in milestone
- New technology
- New personnel
- Lack of training
- New unaligned capability
- Market shift
- Change of management philosophy
- Managers move on
- Depletion of project goodwill
- Lack of user involvement
- Three strikes and out (too many mistakes)

Management by Project

Project management continues to grow as an effective means of managing functions in any organization. Project management should be an enterprise-wide systems-based endeavor. Enterprise-wide project management is the application of project management techniques and practices across the full scope of the enterprise. This concept is also referred to as management by project (MBP). Management by project is an approach that employs project management techniques in various functions within an organization. MBP recommends pursuing endeavors as project-oriented activities. It is an effective way to conduct any business activity. It represents a disciplined approach that defines any work assignment as a project. Under MBP, every undertaking is viewed as a project that must be managed just like a traditional project. The characteristics required of each project so defined are as follows:

1. An identified scope and a goal
2. A desired completion time
3. Availability of resources
4. A defined performance measure
5. A measurement scale for review of work

An MBP approach to operations helps in identifying unique entities within functional requirements. This identification helps determine where functions overlap and how they are interrelated, thus paving the way for better planning, scheduling, and control. Enterprise-wide project management facilitates a unified view of organizational goals and provides a way for project teams to use information generated by other departments to carry out their functions.

The use of project management continues to grow rapidly. The need to develop effective management tools increases with the increasing complexity of new technologies and processes. The life cycle of a new product to be introduced into a competitive market is a good example of a complex process that must be managed with integrative project management approaches. The product will encounter management functions as it goes from one stage to the next. Project management will be needed throughout the design and production stages of the product. Project management will be needed in developing marketing, transportation, and delivery strategies for the product. When the product finally gets to the customer, project management will be needed to integrate its use with those of other products within the customer's organization.

The need for a project management approach is established by the fact that a project will always tend to increase in size even if its scope is narrowing. The following four literary laws are applicable to any project environment:

Parkinson's Law: Work expands to fill the available time or space.

Peter's Principle: People rise to the level of their incompetence.

Murphy's Law: Whatever can go wrong will go wrong.

Badiru's Rule: The grass is always greener where you most need it to be dead.

An integrated systems project management approach can help diminish the adverse impacts of these laws through good project planning, organizing, scheduling, and control.

Integrated Project Implementation

Project management tools can be classified into three major categories:

1. Qualitative tools. There are the managerial tools that aid in the interpersonal and organizational processes required for project management.
2. Quantitative tools. These are analytical techniques that aid in the computational aspects of project management.
3. Computer tools. These are software and hardware tools that simplify the process of planning, organizing, scheduling, and controlling a project. Software tools can help in both the qualitative and quantitative analyses needed for project management.

Although individual books dealing with management principles, optimization models, and computer tools are available, there are few guidelines for the integration of the three areas for project management purposes. In this book, we integrate these three areas for a comprehensive guide to project management. The book introduces the *Triad Approach* to improve the effectiveness of project management with respect to schedule, cost, and performance constraints within the context of systems modeling. Figure 8.5 illustrates this emphasis. The approach considers not only the management of the project itself but also the management of all the functions that support the project. It is

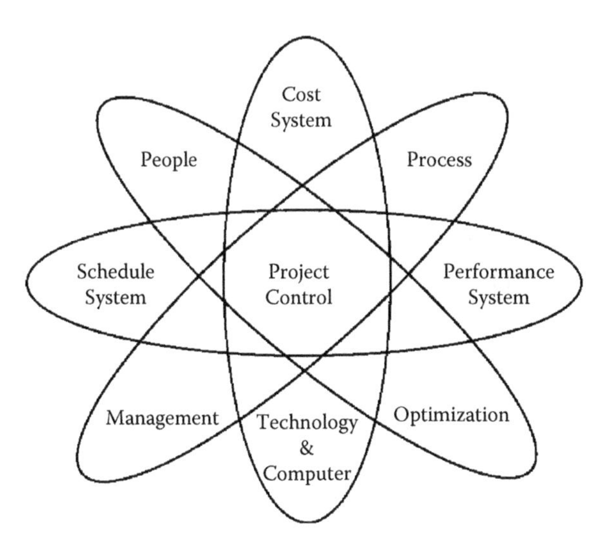

FIGURE 8.5
Project systems view encompassing all aspects of a project.

one thing to have a quantitative model, but it is a different thing to be able to apply the model to real-world problems in a practical form.

A systems approach helps increase the intersection of the three categories of project management tools and, hence, improve overall management effectiveness. Crisis should not be the instigator for the use of project management techniques. Project management approaches should be used up front to prevent avoidable problems rather than to fight them when they develop. What is worth doing is worth doing well, right from the beginning.

Critical Factors for Project Success

The premise of this book is that the critical factors for systems success revolve around people and the personal commitment and dedication of each person. No matter how good a technology is and no matter how enhanced a process might be, it is ultimately the people involved that determine success. This makes it imperative to take care of people issues first in the overall systems approach to project management. Many organizations recognize this, but only a few have been able to actualize the ideals of managing people productively. Execution of operational strategies requires forthrightness, openness, and commitment to get things done. Lip service and arm waving are not sufficient. Tangible programs that cater to the needs of people must be implemented. It is essential to provide incentives, encouragement, and empowerment for people to be self-actuating in determining how best to accomplish their job functions. A summary of critical factors for systems success encompasses the following:

- Total system management: hardware, software, and people
- Operational effectiveness
- Operational efficiency
- System suitability
- System resilience
- System affordability
- System supportability
- System life-cycle cost
- System performance
- System schedule
- System cost

Systems engineering tools, techniques, and processes are essential for project life-cycle management to make goals possible within the context of **SMART** principles, which are represented as follows:

1. Specific: Pursue specific and explicit outputs.
2. Measurable: Design of outputs that can be tracked, measured, and assessed.
3. Achievable: Make outputs to be achievable and aligned with organizational goals.
4. Realistic: Pursue only the goals that are realistic and result-oriented.
5. Timed: Make outputs timed to facilitate accountability.

Project Management Body of Knowledge

The general body of knowledge (PMBOK®) for project management is published and disseminated by the project management institute (PMI®). The body of knowledge comprises specific knowledge areas, which are organized into the following broad areas:

1. Project **Integration** Management
2. Project **Scope** Management
3. Project **Time** Management
4. Project **Cost** Management
5. Project **Quality** Management
6. Project **Human** Resource Management
7. Project **Communications** Management
8. Project **Risk** Management
9. Project **Procurement** Management

The above segments of the body of knowledge of project management cover the range of functions associated with any project, particularly complex ones. Multinational projects particularly pose unique challenges pertaining to reliable power supply, efficient communication systems, credible government support, dependable procurement processes, consistent availability of technology, progressive industrial climate, trustworthy risk mitigation infrastructure, regular supply of skilled labor, uniform focus on quality of work, global consciousness, hassle-free bureaucratic processes, coherent safety and security system, steady law and order, unflinching focus on customer satisfaction, and fair labor relations. Assessing and resolving concerns about these issues in a step-by-step fashion will create a foundation of success for a large project. While no system can be perfect and satisfactory in all aspects, a tolerable trade-off on the factors is essential for project success.

Components of the Knowledge Areas

The key components of each element of the body of knowledge are summarized below:

- Integration
    - Integrative project charter
    - Project scope statement
    - Project management plan
    - Project execution management
    - Change control
- Scope management
    - Focused scope statements
    - Cost/benefits analysis
    - Project constraints
    - Work breakdown structure
    - Responsibility breakdown structure
    - Change control
- Time management
    - Schedule planning and control
    - PERT and Gantt charts
    - Critical Path Method
    - Network models
    - Resource loading
    - Reporting
- Cost management
    - Financial analysis
    - Cost estimating
    - Forecasting
    - Cost control
    - Cost reporting

- Quality management
    - Total quality management
    - Quality assurance
    - Quality control
    - Cost of quality
    - Quality conformance
- Human resources management
    - Leadership skill development
    - Team building
    - Motivation
    - Conflict management
    - Compensation
    - Organizational structures
- Communications
    - Communication matrix
    - Communication vehicles
    - Listening and presenting skills
    - Communication barriers and facilitators
- Risk management
    - Risk identification
    - Risk analysis
    - Risk mitigation
    - Contingency planning
- Procurement and subcontracts
    - Material selection
    - Vendor prequalification
    - Contract types
    - Contract risk assessment
    - Contract negotiation
    - Contract change orders

Step-by-Step and Component-by-Component Implementation

The efficacy of the systems approach is based on step-by-step and component-by-component implementation of the project management process. The major knowledge areas of project management are administered in a structured outline covering six basic clusters consisting of the following:

1. Initiating
2. Planning
3. Executing
4. Monitoring
5. Controlling
6. Closing

The implementation clusters represent five process groups that are followed throughout the project life cycle. Each cluster itself consists of several functions and operational steps. When the clusters are overlaid on the nine knowledge areas, we obtain a two-dimensional matrix that spans 44 major process steps. Table 8.1 shows an overlay of the project management knowledge areas and the implementation clusters. The monitoring and controlling clusters are usually administered as one lumped process group (monitoring and controlling). In some cases, it may be helpful to separate them to highlight the essential attributes of each cluster of functions over the project life cycle. In practice, the processes and clusters do overlap. Thus, there is no crisp demarcation of when and where one process ends and where another one begins over the project life cycle. In general, project life cycle defines the following:

1. Resources that will be needed in each phase of the project life cycle
2. Specific work to be accomplished in each phase of the project life cycle

Figure 8.6 shows the major phases of project life cycle going from the conceptual phase through the closeout phase. It should be noted that project life cycle is distinguished from product life cycle. Project life cycle does not explicitly address operational issues whereas product life cycle is mostly about operational issues starting from the product's delivery to the end of its useful life. Note that for STEP projects, the shape of the life-cycle curve may be expedited due to the rapid developments that often occur in STE activities. For example, for a high technology project, the entire life cycle may be shortened, with a very rapid initial phase, even though the conceptualization stage may be very long. Typical characteristics of project life cycle include the following:

1. Cost and staffing requirements are lowest at the beginning of the project and ramp up during the initial and development stages.
2. The probability of successfully completing the project is lowest at the beginning and highest at the end. This is because many unknowns (risks and uncertainties) exist at the beginning of the project. As the project nears its end, there are fewer opportunities for risks and uncertainties.

TABLE 8.1

Overlay of Project Management Areas and Implementation Clusters

Knowledge Areas	Project Management Process Clusters				
	Initiating	**Planning**	**Executing**	**Monitoring and Controlling**	**Closing**
Project Integration	Develop project charter Develop preliminary project scope	Develop project management plan	Direct and manage project execution	Monitor and control project work Integrated change control	
Scope		Scope planning Scope definition Create WBS		Scope verification Scope control	
Time		Activity definition Activity sequencing Activity resource estimating Activity duration estimating Schedule development	Schedule control		
Cost		Cost estimating Cost budgeting		Cost control	
Quality		Quality planning	Perform quality assurance	Perform quality control	
Human Resources		Human resource planning	Acquire project team Develop project team	Manage project team	
Communication		Communication planning	Information distribution	Performance reporting Manage stakeholders	
Risk		Risk management planning Risk identification Qualitative risk analysis Quantitative risk analysis Risk response planning		Risk monitoring and control	
Procurement		Plan purchases and acquisitions Plan contracting	Request seller responses Select sellers	Contract administration	Contract closure

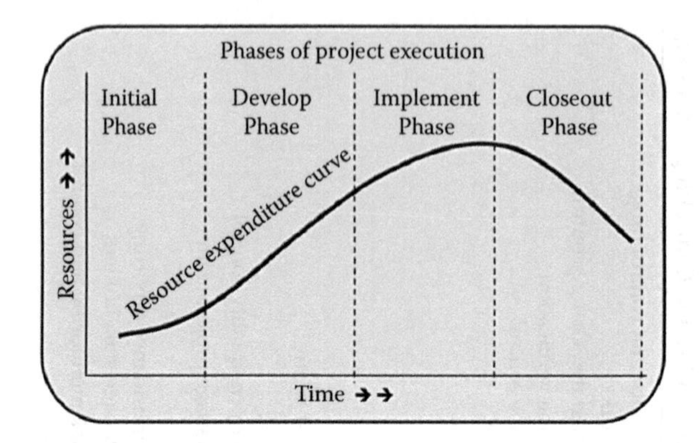

FIGURE 8.6
Project execution phases for systems implementation.

3. The risks to the project organization (project owner) are lowest at the beginning and highest at the end. This is because not much investment has gone into the project at the beginning, whereas much has been committed by the end of the project. There is a higher sunk cost manifested at the end of the project.
4. The ability of the stakeholders to influence the final project outcome (cost, quality, and schedule) is highest at the beginning and gets progressively lower toward the end of the project. This is intuitive because influence is best exerted at the beginning of an endeavor.
5. Value of scope changes decreases over time during the project life cycle while the cost of scope changes increases over time. The suggestion is to decide and finalize scope as early as possible. If there are to be scope changes, do them as early as possible.

Project Systems Structure

The overall project management systems execution can be outlined as summarized below.

Problem Identification

Problem identification is the stage where a need for a proposed project is identified, defined, and justified. A project may be concerned with the development of new products, implementation of new processes, or improvement of existing facilities.

Project Definition

Project definition is the phase at which the purpose of the project is clarified. A *mission statement* is the major output of this stage. For example, a prevailing low level of productivity may indicate a need for a new manufacturing technology. In general, the definition should specify how project management may be used to avoid missed deadlines, poor scheduling, inadequate resource allocation, lack of coordination, poor quality, and conflicting priorities.

Project Planning

A plan represents the outline of the series of actions needed to accomplish a goal. Project planning determines how to initiate a project and execute its objectives. It may be a simple statement of a project goal or it may be a detailed account of procedures to be followed during the project. Project planning is discussed in detail in Chapter 2. Planning can be summarized as

- Objectives
- Project definition
- Team organization
- Performance criteria (time, cost, quality)

Project Organizing

Project organization specifies how to integrate the functions of the personnel involved in a project. Organizing is usually done concurrently with project planning. Directing is an important aspect of project organization. Directing involves guiding and supervising the project personnel. It is a crucial aspect of the management function. Directing requires skillful managers who can interact with subordinates effectively through good communication and motivational techniques. A good project manager will facilitate project success by directing his or her staff, through proper task assignments, toward the project goal.

Workers perform better when there are clearly defined expectations. They need to know how their job functions contribute to the overall goals of the project. Workers should be given some flexibility for self-direction in performing their functions. Individual worker needs and limitations should be recognized by the manager when directing project functions. Directing a project requires skills dealing with motivating, supervising, and delegating.

Resource Allocation

Project goals and objectives are accomplished by allocating resources to functional requirements. Resources can consist of money, people, equipment, tools, facilities, information, skills, and so on. These are usually in short supply. The people needed for a particular task may be committed to other ongoing projects. A crucial piece of equipment may be under the control of another team. Chapter 5 addresses resource allocation in detail.

Project Scheduling

Timeliness is the essence of project management. Scheduling is often the major focus in project management. The main purpose of scheduling is to allocate resources so that the overall project objectives are achieved within a reasonable time span. Project objectives are generally conflicting in nature. For example, minimization of the project completion time and minimization of the project cost are conflicting objectives. That is, one objective is improved at the expense of worsening the other objective. Therefore, project scheduling is a multiple-objective decision-making problem.

In general, scheduling involves the assignment of time periods to specific tasks within the work schedule. Resource availability, time limitations, urgency level, required performance level, precedence requirements, work priorities, technical constraints, and other factors complicate the scheduling process. Thus, the assignment of a time slot to a task does not necessarily ensure that the task will be performed satisfactorily in accordance with the schedule. Consequently, careful control must be developed and maintained throughout the project scheduling process.

Project Tracking and Reporting

This phase involves checking whether project results conform to project plans and performance specifications. Tracking and reporting are prerequisites for project control. A properly organized report of the project status will help identify any deficiencies in the progress of the project and help pinpoint corrective actions.

Project Control

Project control requires that appropriate actions be taken to correct unacceptable deviations from expected performance. Control is actuated through measurement, evaluation, and corrective action. Measurement is the process of measuring the relationship between planned performance and actual performance with respect to project objectives. The variables to be measured, the measurement scales, and the measuring approaches should be clearly specified during the planning stage. Corrective actions may involve rescheduling, reallocation of resources, or expedition of task performance. Project control is discussed in detail in Chapter 6. Control involves the following:

- Tracking and reporting
- Measurement and evaluation
- Corrective action (plan revision, rescheduling, updating)

Project Termination

Termination is the last stage of a project. The phaseout of a project is as important as its initiation. The termination of a project should be implemented expeditiously. A project should not be allowed to drag on after the expected completion time. A terminal activity should be defined for a project during the planning phase. An example of a terminal activity may be the submission of a final report, the power-on of new equipment, or the signing of a release order. The conclusion of such an activity should be viewed as the completion of the project. Arrangements may be made for follow-up activities that may improve or extend the outcome of the project. These follow-up or spinoff projects should be managed as new projects but with proper input-output relationships within the sequence of projects.

Project Systems Implementation Outline

While this book is aligned with the main tenets of PMI's PMBOK, the book uses the traditional project management textbook framework encompassing the broad sequence of categories below:

Planning → Organizing → Scheduling → Control → Termination

An outline of the functions to be carried out during a project should be made during the planning stage of the project. A model for such an outline is presented below. It may be necessary to rearrange the contents of the outline to fit the specific needs of a project.

Planning

I. Specify project background
   A. Define current situation and process
      1. Understand the process
      2. Identify important variables
      3. Quantify variables

B. Identify areas for improvement
   1. List and discuss the areas
   2. Study potential strategy for solution

II. Define unique terminologies relevant to the project
   A. Industry-specific terminologies
   B. Company-specific terminologies
   C. Project-specific terminologies

III. Define project goal and objectives
   A. Write mission statement
   B. Solicit inputs and ideas from personnel

IV. Establish performance standards
   A. Schedule
   B. Performance
   C. Cost

V. Conduct formal project feasibility study
   A. Determine impact on cost
   B. Determine impact on organization
   C. Determine project deliverables

VI. Secure management support

Organizing

I. Identify project management team
   A. Specify project organization structure
      1. Matrix structure
      2. Formal and informal structures
      3. Justify structure
   B. Specify departments involved and key personnel
      1. Purchasing
      2. Materials management
      3. Engineering, design, manufacturing, and so on
   C. Define project management responsibilities
      1. Select project manager
      2. Write project charter
      3. Establish project policies and procedures

II. Implement Triple C Model
- A. Communication
  1. Determine communication interfaces
  2. Develop communication matrix
- B. Cooperation
  1. Outline cooperation requirements, policies, and procedures
- C. Coordination
  1. Develop work breakdown structure
  2. Assign task responsibilities
  3. Develop responsibility chart

Scheduling (Resource Allocation)

I. Develop master schedule
- A. Estimate task duration
- B. Identify task precedence requirements
  1. Technical precedence
  2. Resource-imposed precedence
  3. Procedural precedence
- C. Use analytical models
  1. CPM (Critical Path Method)
  2. PERT (Program Evaluation and Review Technique)
  3. Gantt chart
  4. Optimization models

Control (Tracking, Reporting, and Correction)

I. Establish guidelines for tracking, reporting, and control
- A. Define data requirements
  1. Data categories
  2. Data characterization
  3. Measurement scales
- B. Develop data documentation
  1. Data update requirements
  2. Data quality control
  3. Establish data security measures

II. Categorize control points
  A. Schedule audit
    1. Activity network and Gantt charts
    2. Milestones
    3. Delivery schedule
  B. Performance audit
    1. Employee performance
    2. Product quality
  C. Cost audit
    1. Cost containment measures
    2. Percent completion versus budget depletion

III. Identify implementation process
  A. Comparison with targeted schedules
  B. Corrective course of action
    1. Rescheduling
    2. Reallocation of resources

Termination (Close, Phaseout)

I. Conduct performance review
II. Develop strategy for follow-up projects
III. Arrange for personnel retention, release, and reassignment

Documentation

I. Document project outcome
II. Submit final report
III. Archive report for future reference

In the context of implementing the Triple C Approach to sustainability, the following functions should be addressed.

Sustainability Communication

Many technologies are just emerging from research laboratories. There are still apprehensions and controversies regarding their potential impacts. Implementing new technology projects may generate concerns both within and outside an organization. A frequent concern is the loss of jobs. Sometimes, there may be uncertainties about the impacts of the proposed technology. Proper communication can help educate all the audience of the project concerning its merits. Informative communication is especially important in cases where cultural aspects may influence the success of technology transfer. The people that will be affected by the project should be informed early as to the following:

- The need for the sustainability
- The direct and indirect benefits of sustainability
- The resources that are available to support the technology
- The nature, scope, and the expected impact of the technology
- The expected contributions of individuals involved in the technology
- The person, group, or organization responsible for the technology
- The observers, beneficiaries, and proponents of the technology
- The potential effect of a of the failure of the project
- The funding source for the project

Wide communication is a vital factor in securing support for sustainability. A concerted effort should be made to inform those who should know. Moreover, the communication channel must be kept open throughout the project life cycle. In addition to in-house communication, external sources should also be consulted as appropriate. A sustainability consortium may be established to facilitate communication with external sources. The consortium will link various organizations with respect to specific technology products and objectives. This will facilitate exchange of both technical and managerial ideas.

Sustainability Cooperation

Not only must people be informed, but their cooperation must also be explicitly sought. Merely saying "yeah" to sustainability is not enough assurance of full cooperation. In effect, the proposed technology must be sold to management and employees. A structured approach to seeking cooperation should help identify and explain the following items to the project personnel:

The cooperative efforts needed to ensure success of the technology
The time frame involved in implementing the technology
The criticality of cooperation to the technology
The organizational benefits of cooperation
The implication of lack of cooperation

Sustainability Coordination

Having successfully initiated the communication and cooperation functions, the efforts of the project team must, thereafter, be coordinated. Coordination facilitates the organization and utilization of resources. The development of a responsibility chart can be very helpful at this stage. A responsibility chart is a matrix consisting of columns of individual or functional departments and rows of required actions. Cells within the matrix are filled with relationship codes that indicate who is responsible for what. The matrix should indicate the following:

- Who is to do what
- Who is to inform whom of what
- Whose approval is needed for what
- Who is responsible for which results
- What personnel interfaces are involved
- What support is needed from whom for what functions

The use of a project management approach is particularly important when sustainability technology is transferred from a developed nation (or organization) to a less developed nation (or organization). In some cases, fully completed technology products cannot be transferred due to the incompatibility of operating conditions and requirements. In some cases, the receiving organization has the means to adapt transferred technology concepts, theories, and ideas to local conditions to generate the desired products. In other cases, the receiving organization has the infrastructure to implement technology procedures and guidelines to obtain the required products at the local level. To reach the overall goal of sustainability technology transfer, it is essential that the most suitable technology be identified promptly, transferred under the most favorable terms, and implemented at the receiving organization in the most appropriate manner. Project management offers guidelines and models that can be helpful in achieving these aims.

Project Decision Analysis

Systems decision analysis facilitates a proper consideration of the essential elements of decisions in a project systems environment. These essential elements include the problem statement, information, performance measure, decision model, and an implementation of the decision. The recommended steps are enumerated below.

Step 1: Problem Statement

A problem involves choosing between competing, and probably conflicting, alternatives. The components of problem solving in project management include the following:

- Describing the problem (goals, performance measures)
- Defining a model to represent the problem
- Solving the model
- Testing the solution
- Implementing and maintaining the solution

Problem definition is very crucial. In many cases, *symptoms* of a problem are more readily recognized than its *cause* and *location*. Even after the problem is accurately identified and defined, a benefit/cost analysis may be needed to determine if the cost of solving the problem is justified.

Step 2: Data and Information Requirements

Information is the driving force for the project decision process. Information clarifies the relative states of past, present, and future events. The collection, storage, retrieval, organization, and processing of raw data are important components for generating information. Without data, there can be no information. Without good information, there cannot be a valid decision. The essential requirements for generating information are as follows:

- Ensuring that an effective data collection procedure is followed
- Determining the type and the appropriate amount of data to collect
- Evaluating the data collected with respect to information potential
- Evaluating the cost of collecting the required data

For example, suppose a manager is presented with a recorded fact that says, "Sales for the last quarter are 10,000 units." This constitutes ordinary data. There are many ways of using the above data to make a decision depending on the manager's value system. An analyst, however, can ensure the proper use of the data by transforming it into information, such as, "Sales of 10,000 units for last quarter are within *x* percent of the targeted value." This type of information is more useful to the manager for decision making.

Step 3: Performance Measure

A performance measure for the competing alternatives should be specified. The decision maker assigns a perceived worth or value to the available alternatives. Setting measures of performance is crucial to the process of defining and selecting alternatives. Some performance measures commonly used in project management are project cost, completion time, resource usage, and stability in the workforce.

Step 4: Decision Model

A decision model provides the basis for the analysis and synthesis of information and is the mechanism by which competing alternatives are compared. To be effective, a decision model must be based on a systematic and logical framework for guiding project decisions. A decision model can be a verbal, graphical, or mathematical representation of the ideas in the decision-making process. A project decision model should have the following characteristics:

- Simplified representation of the actual situation
- Explanation and prediction of the actual situation
- Validity and appropriateness
- Applicability to similar problems

The formulation of a decision model involves three essential components:

Abstraction: Determining the relevant factors
Construction: Combining the factors into a logical model
Validation: Ensuring that the model adequately represents the problem

The basic types of decision models for project management are described next.

Descriptive models. These models are directed at describing a decision scenario and identifying the associated problem. For example, a project analyst might use a critical path method (CPM) network model to identify bottleneck tasks in a project.

Prescriptive models. These models furnish procedural guidelines for implementing actions. The Triple C approach (Badiru, 2008), for example, is a model that prescribes the procedures for achieving communication, cooperation, and coordination in a project environment.

Predictive models. These models are used to predict future events in a problem environment. They are typically based on historical data about the problem situation. For example, a regression model based on past data may be used to predict future productivity gains associated with expected levels of resource allocation. Simulation models can be used when uncertainties exist in the task durations or resource requirements.

"Satisficing" models. These are models that provide trade-off strategies for achieving a satisfactory solution to a problem within given constraints. *Satisficing*, a concatenation of the words *satisfy* and *suffice*, is a decision-making strategy that attempts to meet criteria for adequacy, rather than to identify an optimal solution. This is used where achieving an optimum is not practicable. Goal programming and other multicriteria techniques provide good satisficing solutions. For example, these models are helpful in cases where time limitations, resource shortages, and performance requirements constrain the implementation of a project.

Optimization models. These models are designed to find the best available solution to a problem subject to a certain set of constraints. For example, a linear programming model can be used to determine the optimal product mix in a production environment.

In many situations, two or more of the above models may be involved in the solution of a problem. For example, a descriptive model might provide insights into the nature of the problem; an optimization model might provide the optimal set of actions to take in solving the problem; a satisficing model might temper the optimal solution with reality; a prescriptive model might suggest the procedures for implementing the selected solution; and a predictive model is created from predictions of the most likely probability of an outcome.

Step 5: Making the Decision

Using the available data, information, and the decision model, the decision maker will determine the real-world actions that are needed to solve the stated problem. A sensitivity analysis may be useful for determining what changes in parameter values might cause a change in the decision.

Step 6: Implementing the Decision

A decision represents the selection of an alternative that satisfies the objective stated in the problem statement. A good decision is useless until it is implemented. An important aspect of a decision is to specify how it is to be implemented. Selling the decision and the project to management requires a well-organized persuasive presentation. The way a decision is presented can directly influence whether or not it is adopted. The presentation of a decision should include at least the following: an executive summary, technical aspects of the decision, managerial aspects of the decision, resources required to implement the decision, cost of the decision, the time frame for implementing the decision, and the risks associated with the decision.

Systems Group Decision-Making Models

Systems decisions are often complex, diffuse, distributed, and poorly understood. No one person has all the information to make all decisions accurately.

As a result, crucial decisions are made by a group of people. Some organizations use outside consultants with appropriate expertise to make recommendations for important decisions. Other organizations set up their own internal consulting groups without having to go outside the organization. Decisions can be made through linear responsibility, in which case one person makes the final decision based on inputs from other people. Decisions can also be made through shared responsibility, in which case a group of people share the responsibility for making joint decisions. The major advantages of group decision making are listed below:

1. Facilitation of a systems view of the problem environment.
2. Ability to share experience, knowledge, and resources. Many heads are better than one. A group will possess greater collective ability to solve a given decision problem.
3. Increased credibility. Decisions made by a group of people often carry more weight in an organization.
4. Improved morale. Personnel morale can be positively influenced because many people have the opportunity to participate in the decision-making process.
5. Better rationalization. The opportunity to observe other people's views can lead to an improvement in an individual's reasoning process.
6. Ability to accumulate more knowledge and facts from diverse sources.
7. Access to broader perspectives spanning different problem scenarios.
8. Ability to generate and consider alternatives from different perspectives.
9. Possibility for a broader-based involvement, leading to a higher likelihood of support.
10. Possibility for group leverage for networking, communication, and political clout.

In spite of the much-desired advantages, group decision making does possess the risk of flaws. Some possible disadvantages of group decision making are listed below:

1. Difficulty in arriving at a decision.
2. Slow operating timeframe.
3. Possibility for individuals conflicting views and objectives.
4. Reluctance of some individuals in implementing the decision.
5. Potential for power struggle and conflicts among the group.
6. Loss of productive employee time.
7. Too much compromise may lead to less than optimal group output.
8. Risk of one individual dominating the group.

9. Overreliance on group process may impede agility of management to make decisions fast.
10. Risk of dragging feet due to repeated and iterative group meetings.

Brainstorming

Brainstorming is a way of generating many new ideas. In brainstorming, the decision group comes together to discuss alternate ways of solving a problem. The members of the brainstorming group may be from different departments, may have different backgrounds and training, and may not even know one another. The diversity of the participants helps create a stimulating environment for generating different ideas from different viewpoints. The technique encourages free outward expression of new ideas no matter how far-fetched the ideas might appear. No criticism of any new idea is permitted during the brainstorming session. A major concern in brainstorming is that extroverts may take control of the discussions. For this reason, an experienced and respected individual should manage the brainstorming discussions. The group leader establishes the procedure for proposing ideas, keeps the discussions in line with the group's mission, discourages disruptive statements, and encourages the participation of all members.

After the group runs out of ideas, open discussions are held to weed out the unsuitable ones. It is expected that even the rejected ideas may stimulate the generation of other ideas which may eventually lead to other favored ideas. Guidelines for improving brainstorming sessions are presented as follows:

- Focus on a specific decision problem.
- Keep ideas relevant to the intended decision.
- Be receptive to all new ideas.
- Evaluate the ideas on a relative basis after exhausting new ideas.
- Maintain an atmosphere conducive to cooperative discussions.
- Maintain a record of the ideas generated.

Delphi Method

The traditional approach to group decision making is to obtain the opinion of experienced participants through open discussions. An attempt is made to reach a consensus among the participants. However, open group discussions are often biased because of the influence of subtle intimidation from dominant individuals. Even when the threat of a dominant individual is not present, opinions may still be swayed by group pressure. This is called the "bandwagon effect" of group decision making.

The Delphi method attempts to overcome these difficulties by requiring individuals to present their opinions anonymously through an intermediary.

The method differs from the other interactive group methods because it eliminates face-to-face confrontations. It was originally developed for forecasting applications, but it has been modified in various ways for application to different types of decision making. The method can be quite useful for project management decisions. It is particularly effective when decisions must be based on a broad set of factors. The Delphi method is normally implemented as follows:

1. Problem definition. A decision problem that is considered significant is identified and clearly described.
2. Group selection. An appropriate group of experts or experienced individuals is formed to address the particular decision problem. Both internal and external experts may be involved in the Delphi process. A leading individual is appointed to serve as the administrator of the decision process. The group may operate through the mail or gather together in a room. In either case, all opinions are expressed anonymously on paper. If the group meets in the same room, care should be taken to provide enough room so that each member does not have the feeling that someone may accidentally or deliberately observe their responses.
3. Initial opinion poll. The technique is initiated by describing the problem to be addressed in unambiguous terms. The group members are requested to submit a list of major areas of concern in their specialty areas as they relate to the decision problem.
4. Questionnaire design and distribution. Questionnaires are prepared to address the areas of concern related to the decision problem. The written responses to the questionnaires are collected and organized by the administrator. The administrator aggregates the responses in a statistical format. For example, the average, mode, and median of the responses may be computed. This analysis is distributed to the decision group. Each member can then see how his or her responses compare with the anonymous views of the other members.
5. Iterative balloting. Additional questionnaires based on the previous responses are passed to the members. The members submit their responses again. They may choose to alter or not to alter their previous responses.
6. Silent discussions and consensus. The iterative balloting may involve anonymous written discussions of why some responses are correct or incorrect. The process is continued until a consensus is reached. A consensus may be declared after five or six iterations of the balloting or when a specified percentage (e.g., 80%) of the group agrees on the questionnaires. If a consensus cannot be declared on a particular

point, it may be displayed to the whole group with a note that it does not represent a consensus.

In addition to its use in technological forecasting, the Delphi method has been widely used in other general decision making. Its major characteristics of anonymity of responses, statistical summary of responses, and controlled procedure make it a reliable mechanism for obtaining numeric data from subjective opinion. The major limitations of the Delphi method are as follows:

1. Its effectiveness may be limited in cultures where strict hierarchy, seniority, and age influence decision-making processes.
2. Some experts may not readily accept the contribution of non-experts to the group decision-making process.
3. Since opinions are expressed anonymously, some members may take the liberty of making ludicrous statements. However, if the group composition is carefully reviewed, this problem may be avoided.

Nominal Group Technique

The nominal group technique is a silent version of brainstorming. It is a method of reaching consensus. Rather than asking people to state their ideas aloud, the team leader asks each member to jot down a minimum number of ideas, for example, five or six. A single list of ideas is then written on a chalkboard for the whole group to see. The group then discusses the ideas and weeds out some iteratively until a final decision is made. The nominal group technique is easier to control. Unlike brainstorming where members may get into shouting matches, the nominal group technique permits members to silently present their views. In addition, it allows introversive members to contribute to the decision without the pressure of having to speak out too often.

In all of the group decision-making techniques, an important aspect that can enhance and expedite the decision-making process is to require that members review all pertinent data before coming to the group meeting. This will ensure that the decision process is not impeded by trivial preliminary discussions. Some disadvantages of group decision making are as follows:

1. Peer pressure in a group situation may influence a member's opinion or discussions.
2. In a large group, some members may not get to participate effectively in the discussions.
3. A member's relative reputation in the group may influence how well his or her opinion is rated.

4. A member with a dominant personality may overwhelm the other members in the discussions.
5. The limited time available to the group may create a time pressure that forces some members to present their opinions without fully evaluating the ramifications of the available data.
6. It is often difficult to get all members of a decision group together at the same time.

Despite the noted disadvantages, group decision making definitely has many advantages that may nullify the shortcomings. The advantages as presented earlier will have varying levels of effect from one organization to another. The Triple C principle introduced in Chapter 6 may also be used to improve the success of decision teams. Teamwork can be enhanced in group decision making by adhering to the following guidelines:

1. Get a willing group of people together.
2. Set an achievable goal for the group.
3. Determine the limitations of the group.
4. Develop a set of guiding rules for the group.
5. Create an atmosphere conducive to group synergism.
6. Identify the questions to be addressed in advance.
7. Plan to address only one topic per meeting.

For major decisions and long-term group activities, arrange for team training that allows the group to learn the decision rules and responsibilities together. The steps for the nominal group technique are as follows:

1. Silently generate ideas, in writing.
2. Record ideas without discussion.
3. Conduct group discussion for clarification of meaning, not argument.
4. Vote to establish the priority or rank of each item.
5. Discuss vote.
6. Cast final vote.

Interviews, Surveys, and Questionnaires

Interviews, surveys, and questionnaires are important information-gathering techniques. They also foster cooperative working relationships. They encourage direct participation and inputs into project decision-making processes. They provide an opportunity for employees at the lower levels of an organization to contribute ideas and inputs for decision making. The greater the number of people involved in the interviews, surveys, and questionnaires, the more valid the final decision. The following guidelines are useful for conducting interviews, surveys, and questionnaires to collect data and information for project decisions:

1. Collect and organize background information and supporting documents on the items to be covered by the interview, survey, or questionnaire.
2. Outline the items to be covered and list the major questions to be asked.
3. Use a suitable medium of interaction and communication: telephone, fax, electronic mail, face-to-face, observation, meeting venue, poster, or memo.
4. Tell the respondent the purpose of the interview, survey, or questionnaire, and indicate how long it will take.
5. Use open-ended questions that stimulate ideas from the respondents.
6. Minimize the use of yes or no type of questions.
7. Encourage expressive statements that indicate the respondent's views.
8. Use the who, what, where, when, why, and how approach to elicit specific information.
9. Thank the respondents for their participation.
10. Let the respondents know the outcome of the exercise.

Multivote

Multivoting is a series of votes used to arrive at a group decision. It can be used to assign priorities to a list of items. It can be used at team meetings after a brainstorming session has generated a long list of items. Multivoting helps reduce such long lists to a few items, usually three to five. The steps for multivoting are listed below:

1. Take a first vote. Each person votes as many times as desired, but only once per item.
2. Circle the items receiving a relatively higher number of votes (i.e., majority vote) than the other items.
3. Take a second vote. Each person votes for a number of items equal to one-half the total number of items circled in step 2. Only one vote per item is permitted.
4. Repeat steps 2 and 3 until the list is reduced to three to five items depending on the needs of the group. It is not recommended to multivote down to only one item.
5. Perform further analysis of the items selected in step 4, if needed.

Hierarchy of Project Control

The traditional concepts of systems analysis are applicable to the project process. The definitions of a project system and its components are presented next from a point-to-point control perspective.

- **System.** A project system consists of interrelated elements organized for the purpose of achieving a common goal. The elements are organized to work synergistically to generate a unified output that is greater than the sum of the individual outputs of the components.
- **Program.** A program is a very large and prolonged undertaking. Such endeavors often span several years. Programs are usually associated with particular systems. For example, we may have a space exploration program within a national defense system.
- **Project.** A project is a time-phased effort of much smaller scope and duration than a program. Programs are sometimes viewed as consisting of a set of projects. Government projects are often called *programs* because of their broad and comprehensive nature. Industry tends to use the term *project* because of the short-term and focused nature of most industrial efforts.
- **Task.** A task is a functional element of a project. A project is composed of a sequence of tasks that all contribute to the overall project goal.
- **Activity.** An activity can be defined as a single element of a project. Activities are generally smaller in scope than tasks. In a detailed analysis of a project, an activity may be viewed as the smallest, practically indivisible work element of the project. For example, we can regard a manufacturing plant as a system. A plantwide endeavor to improve productivity can be viewed as a program. The installation of a flexible manufacturing system is a project within the productivity improvement program. The process of identifying and selecting equipment vendors is a task, and the actual process of placing an order with a preferred vendor is an activity. The systems structure of a project is illustrated in Figure 8.7.

The emergence of systems development has had an extensive effect on project management in recent years. A system can be defined as a collection of interrelated elements brought together to achieve a specified objective. In a management context, the purposes of a system are to develop and manage operational procedures and to facilitate an effective decision-making process. Some of the common characteristics of a system include the following:

FIGURE 8.7
Hierarchy of a project system.

1. Interaction with the environment
2. Objective
3. Self-regulation
4. Self-adjustment

Representative components of a project system are the organizational subsystem, planning subsystem, scheduling subsystem, information management subsystem, control subsystem, and project delivery subsystem. The primary responsibilities of project analysts involve ensuring the proper flow of information throughout the project system. The classical approach to the decision process follows rigid lines of organizational charts. By contrast, the systems approach considers all the interactions necessary among the various elements of an organization in the decision process.

The various elements (or subsystems) of the organization act simultaneously in a separate but interrelated fashion to achieve a common goal. This synergism helps to expedite the decision process and to enhance the effectiveness of decisions. The supporting commitments from other subsystems of the organization serve to counterbalance the weaknesses of a given subsystem. Thus, the overall effectiveness of the system is greater than the sum of the individual results from the subsystems.

The increasing complexity of organizations and projects makes the systems approach essential in today's management environment. As the number of complex projects increases, there will be an increasing need for project management professionals who can function as systems integrators.

Project management techniques can be applied to the various stages of implementing a system as shown in the following guidelines:

1. Systems definition. Define the system and associated problems using keywords that signify the importance of the problem to the overall organization. Locate experts in this area who are willing to contribute to the effort. Prepare and announce the development plan.
2. Personnel assignment. The project group and the respective tasks should be announced, a qualified project manager should be appointed, and a solid line of command should be established and enforced.
3. Project initiation. Arrange an organizational meeting during which a general approach to the problem should be discussed. Prepare a specific development plan and arrange for the installation of needed hardware and tools.
4. System prototype. Develop a prototype system, test it, and learn more about the problem from the test results.
5. Full system development. Expand the prototype to a full system, evaluate the user interface structure, and incorporate user training facilities and documentation.
6. System verification. Get experts and potential users involved, ensure that the system performs as designed, and debug the system as needed.
7. System validation. Ensure that the system yields expected outputs. Validate the system by evaluating performance level, such as percentage of success in so many trials, measuring the level of deviation from expected outputs, and measuring the effectiveness of the system output in solving the problem.
8. System integration. Implement the full system as planned, ensure that the system can coexist with systems already in operation, and arrange for technology transfer to other projects.
9. System maintenance. Arrange for continuing maintenance of the system. Update solution procedures as new pieces of information become available. Retain responsibility for system performance or delegate to well-trained and authorized personnel.
10. Documentation. Prepare full documentation of the system, prepare a user's guide, and appoint a user consultant. Pictures and graphics are helpful resources for user's guides.

Systems integration permits sharing of resources. Physical equipment, concepts, information, and skills may be shared as resources. Systems integration is now a major concern of many organizations. Even some of the organizations that traditionally compete and typically shun cooperative efforts are beginning to appreciate the value of integrating their operations. For these reasons, systems integration has emerged as a major interest in business. Systems integration may involve the physical integration of technical components, objective integration of operations, conceptual integration of management processes, or a combination of any of these.

Systems integration involves the linking of components to form subsystems and the linking of subsystems to form composite systems within a single department and/or across departments. It facilitates the coordination of technical and managerial efforts to enhance organizational functions, reduce cost, save energy, improve productivity, and increase the utilization of resources. Systems integration emphasizes the identification and coordination of the interface requirements among the components in an integrated system. The components and subsystems operate synergistically to optimize the performance of the total system. Systems integration ensures that all performance goals are satisfied with a minimum expenditure of time and resources. Integration can be achieved in several forms including the following:

1. Dual-use integration: This involves the use of a single component by separate subsystems to reduce both the initial cost and the operating cost during the project life cycle.
2. Dynamic resource integration: This involves integrating resources that flow between two normally separate subsystems of a project, in such a way that resources flow from one subsystem to or through the other subsystem, in a way that minimizes the total resource requirements of the project.
3. Restructuring of functions: This involves the restructuring of functions and reintegration of subsystems to optimize costs when a new subsystem is introduced into the project environment.

Systems integration is particularly important when introducing new technology into an existing system. It involves coordinating new operations to coexist with existing operations. It may require the adjustment of functions to permit the sharing of resources, development of new policies to accommodate product integration, or realignment of managerial responsibilities. It can affect both hardware and software components of an organization. Presented below are guidelines and important questions relevant for systems integration.

- What are the unique characteristics of each component in the integrated system?
- How do the characteristics complement one another?

- What physical interfaces exist among the components?
- What data/information interfaces exist among the components?
- What ideological differences exist among the components?
- What are the data flow requirements for the components?
- Are there similar integrated systems operating elsewhere?
- What are the reporting requirements in the integrated system?
- Are there any hierarchical restrictions on the operations of the components of the integrated system?
- What internal and external factors are expected to influence the integrated system?
- How can the performance of the integrated system be measured?
- What benefit/cost documentations are required for the integrated system?
- What is the cost of designing and implementing the integrated system?
- What are the relative priorities assigned to each component of the integrated system?
- What are the strengths of the integrated system?
- What are the weaknesses of the integrated system?
- What resources are needed to keep the integrated system operating satisfactorily?
- Which section of the organization will have primary responsibility for the operation of the integrated system?
- What are the quality specifications and requirements for the integrated systems?

The integrated approach to project management recommended in this book is represented by the flowchart in Figure 8.8. Figure 8.9 illustrates an example of the application of matrix organization structure in the pursuit of a function goal, where horizontal and vertical lines of responsibilities share knowledge, resources, and personnel.

The process starts with a managerial analysis of the project effort. Goals and objectives are defined, a mission statement is written, and the statement of work is developed. After these, traditional project management approaches, such as the selection of an organization structure, are employed. Conventional analytical tools including the critical path method (CPM) and the precedence diagramming method (PERT) are then mobilized. The use of optimization models is then called upon as appropriate. Some of the parameters to be optimized are cost, resource allocation, and schedule length. It

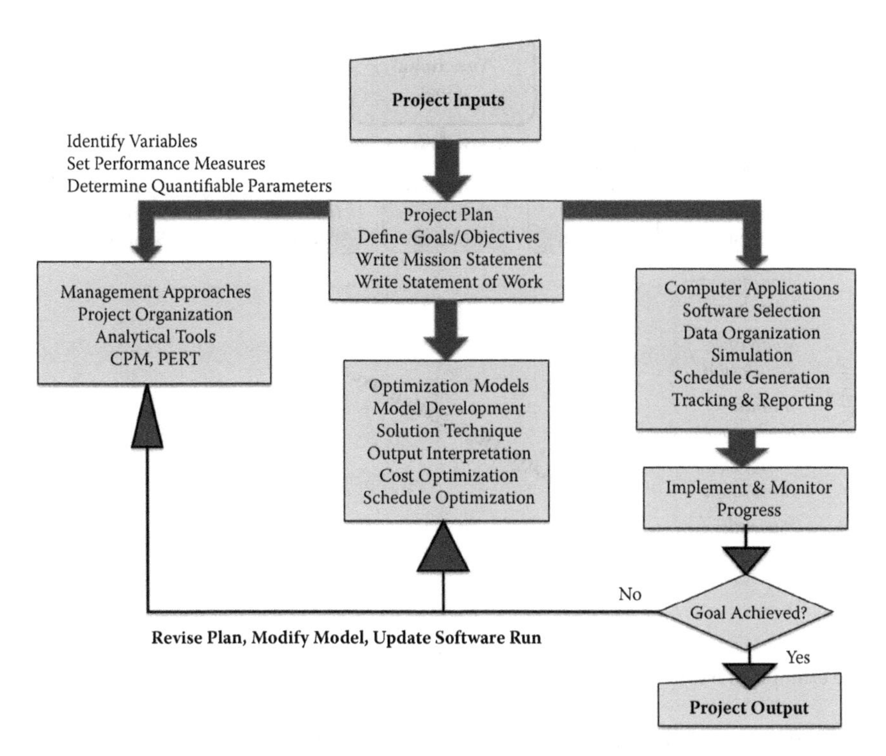

FIGURE 8.8
Flowchart of integrated project management.

should be understood that not all project parameters will be amenable to optimization. The use of commercial project management software should start only after the managerial functions have been completed. Some project management software have built-in capabilities for the planning and optimization needs.

A frequent mistake in project management is the rush to use project management software without first completing the planning and analytical studies required by the project. Project management software should be used as a management tool, the same way a word processor is used as a writing tool. It will not be effective to start using the word processor without first organizing the thoughts about what is to be written. Project management is much more than just the project management software. If project management is carried out in accordance with the integration approach presented in the flowchart, the odds of success will be increased. Of course, the structure of the flowchart should not be rigid. Flows and interfaces among the blocks

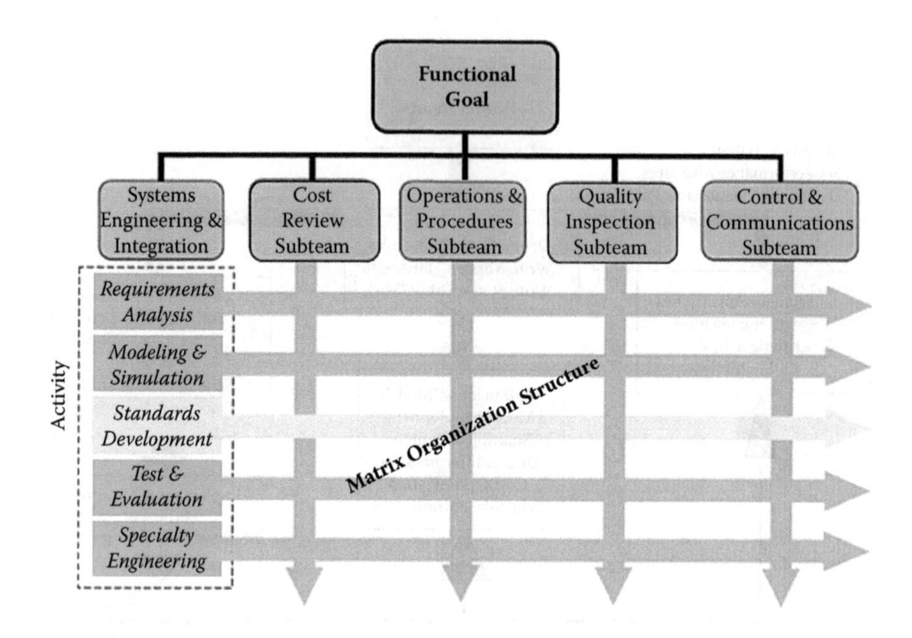

FIGURE 8.9
Case example of matrix organization structure.

in the flowchart may need to be altered or modified depending on specific project needs.

Reference

Badiru, A.B. (2012). *Project Management: Systems, Principles, and Applications*. CRC Press/Taylor and Francis, Boca Raton, FL.

Appendix A: Cumulative Normal Probability Tables (Z-Values)

Z	0.00	0.01	0.02	0.03	0.04	0.05	0.06	0.07	0.08	0.09
0.0	0.50000	0.50399	0.50798	0.51197	0.51595	0.51994	0.52392	0.52790	0.53188	0.53586
0.1	0.53983	0.54380	0.54776	0.55172	0.55567	0.55962	0.56356	0.56749	0.57142	0.57535
0.2	0.57926	0.58317	0.58706	0.59095	0.59483	0.59871	0.60257	0.60642	0.61026	0.61409
0.3	0.61791	0.62172	0.62552	0.62930	0.63307	0.63683	0.64058	0.64431	0.64803	0.65173
0.4	0.65542	0.65910	0.66276	0.66640	0.67003	0.67364	0.67724	0.68082	0.68439	0.68793
0.5	0.69146	0.69497	0.69847	0.70194	0.70540	0.70884	0.71226	0.71566	0.71904	0.72240
0.6	0.72575	0.72907	0.73237	0.73565	0.73891	0.74215	0.74537	0.74857	0.75175	0.75490
0.7	0.75804	0.76115	0.76424	0.76730	0.77035	0.77337	0.77637	0.77935	0.78230	0.78524
0.8	0.78814	0.79103	0.79389	0.79673	0.79955	0.80234	0.80511	0.80785	0.81057	0.81327
0.9	0.81594	0.81859	0.82121	0.82381	0.82639	0.82894	0.83147	0.83398	0.83646	0.83891
1.0	0.84134	0.84375	0.84614	0.84849	0.85083	0.85314	0.85543	0.85769	0.85993	0.86214
1.1	0.86433	0.86650	0.86864	0.87076	0.87286	0.87493	0.87698	0.87900	0.88100	0.88298
1.2	0.88493	0.88686	0.88877	0.89065	0.89251	0.89435	0.89617	0.89796	0.89973	0.90147
1.3	0.90320	0.90490	0.90658	0.90824	0.90988	0.91149	0.91308	0.91466	0.91621	0.91774
1.4	0.91924	0.92073	0.92220	0.92364	0.92507	0.92647	0.92785	0.92922	0.93056	0.93189
1.5	0.93319	0.93448	0.93574	0.93699	0.93822	0.93943	0.94062	0.94179	0.94295	0.94408
1.6	0.94520	0.94630	0.94738	0.94845	0.94950	0.95053	0.95154	0.95254	0.95352	0.95449
1.7	0.95543	0.95637	0.95728	0.95818	0.95907	0.95994	0.96080	0.96164	0.96246	0.96327
1.8	0.96407	0.96485	0.96562	0.96638	0.96712	0.96784	0.96856	0.96926	0.96995	0.97062
1.9	0.97128	0.97193	0.97257	0.97320	0.97381	0.97441	0.97500	0.97558	0.97615	0.97670
2.0	0.97725	0.97778	0.97831	0.97882	0.97932	0.97982	0.98030	0.98077	0.98124	0.98169
2.1	0.98214	0.98257	0.98300	0.98341	0.98382	0.98422	0.98461	0.98500	0.98537	0.98574
2.2	0.98610	0.98645	0.98679	0.98713	0.98745	0.98778	0.98809	0.98840	0.98870	0.98899
2.3	0.98928	0.98956	0.98983	0.99010	0.99036	0.99061	0.99086	0.99111	0.99134	0.99158
2.4	0.99180	0.99202	0.99224	0.99245	0.99266	0.99286	0.99305	0.99324	0.99343	0.99361
2.5	0.99379	0.99396	0.99413	0.99430	0.99446	0.99461	0.99477	0.99492	0.99506	0.99520
2.6	0.99534	0.99547	0.99560	0.99573	0.99585	0.99598	0.99609	0.99621	0.99632	0.99643
2.7	0.99653	0.99664	0.99674	0.99683	0.99693	0.99702	0.99711	0.99720	0.99728	0.99736
2.8	0.99744	0.99752	0.99760	0.99767	0.99774	0.99781	0.99788	0.99795	0.99801	0.99807
2.9	0.99813	0.99819	0.99825	0.99831	0.99836	0.99841	0.99846	0.99851	0.99856	0.99861

continued

Z	0.00	0.01	0.02	0.03	0.04	0.05	0.06	0.07	0.08	0.09
3.0	0.99865	0.99869	0.99874	0.99878	0.99882	0.99886	0.99889	0.99893	0.99896	0.99900
3.1	0.99903	0.99906	0.99910	0.99913	0.99916	0.99918	0.99921	0.99924	0.99926	0.99929
3.2	0.99931	0.99934	0.99936	0.99938	0.99940	0.99942	0.99944	0.99946	0.99948	0.99950
3.3	0.99952	0.99953	0.99955	0.99957	0.99958	0.99960	0.99961	0.99962	0.99964	0.99965
3.4	0.99966	0.99968	0.99969	0.99970	0.99971	0.99972	0.99973	0.99974	0.99975	0.99976
3.5	0.99977	0.99978	0.99978	0.99979	0.99980	0.99981	0.99981	0.99982	0.99983	0.99983
3.6	0.99984	0.99985	0.99985	0.99986	0.99986	0.99987	0.99987	0.99988	0.99988	0.99989
3.7	0.99989	0.99990	0.99990	0.99990	0.99991	0.99991	0.99992	0.99992	0.99992	0.99992
3.8	0.99993	0.99993	0.99993	0.99994	0.99994	0.99994	0.99994	0.99995	0.99995	0.99995
3.9	0.99995	0.99995	0.99996	0.99996	0.99996	0.99996	0.99996	0.99996	0.99997	0.99997
4.0	0.99997	0.99997	0.99997	0.99997	0.99997	0.99997	0.99998	0.99998	0.99998	0.99998
–4.0	0.00002	0.00002	0.00002	0.00002	0.00003	0.00003	0.00003	0.00003	0.00003	0.00003
–3.9	0.00003	0.00003	0.00004	0.00004	0.00004	0.00004	0.00004	0.00004	0.00005	0.00005
–3.8	0.00005	0.00005	0.00005	0.00006	0.00006	0.00006	0.00006	0.00007	0.00007	0.00007
–3.7	0.00008	0.00008	0.00008	0.00008	0.00009	0.00009	0.00010	0.00010	0.00010	0.00011
–3.6	0.00011	0.00012	0.00012	0.00013	0.00013	0.00014	0.00014	0.00015	0.00015	0.00016
–3.5	0.00017	0.00017	0.00018	0.00019	0.00019	0.00020	0.00021	0.00022	0.00022	0.00023
–3.4	0.00024	0.00025	0.00026	0.00027	0.00028	0.00029	0.00030	0.00031	0.00032	0.00034
–3.3	0.00035	0.00036	0.00038	0.00039	0.00040	0.00042	0.00043	0.00045	0.00047	0.00048
–3.2	0.00050	0.00052	0.00054	0.00056	0.00058	0.00060	0.00062	0.00064	0.00066	0.00069
–3.1	0.00071	0.00074	0.00076	0.00079	0.00082	0.00084	0.00087	0.00090	0.00094	0.00097
–3.0	0.00100	0.00104	0.00107	0.00111	0.00114	0.00118	0.00122	0.00126	0.00131	0.00135
–2.9	0.00139	0.00144	0.00149	0.00154	0.00159	0.00164	0.00169	0.00175	0.00181	0.00187
–2.8	0.00193	0.00199	0.00205	0.00212	0.00219	0.00226	0.00233	0.00240	0.00248	0.00256
–2.7	0.00264	0.00272	0.00280	0.00289	0.00298	0.00307	0.00317	0.00326	0.00336	0.00347
–2.6	0.00357	0.00368	0.00379	0.00391	0.00402	0.00415	0.00427	0.00440	0.00453	0.00466
–2.5	0.00480	0.00494	0.00508	0.00523	0.00539	0.00554	0.00570	0.00587	0.00604	0.00621
–2.4	0.00639	0.00657	0.00676	0.00695	0.00714	0.00734	0.00755	0.00776	0.00798	0.00820
–2.3	0.00842	0.00866	0.00889	0.00914	0.00939	0.00964	0.00990	0.01017	0.01044	0.01072
–2.2	0.01101	0.01130	0.01160	0.01191	0.01222	0.01255	0.01287	0.01321	0.01355	0.01390
–2.1	0.01426	0.01463	0.01500	0.01539	0.01578	0.01618	0.01659	0.01700	0.01743	0.01786
–2.0	0.01831	0.01876	0.01923	0.01970	0.02018	0.02068	0.02118	0.02169	0.02222	0.02275
–1.9	0.02330	0.02385	0.02442	0.02500	0.02559	0.02619	0.02680	0.02743	0.02807	0.02872
–1.8	0.02938	0.03005	0.03074	0.03144	0.03216	0.03288	0.03362	0.03438	0.03515	0.03593
–1.7	0.03673	0.03754	0.03836	0.03920	0.04006	0.04093	0.04182	0.04272	0.04363	0.04457
–1.6	0.04551	0.04648	0.04746	0.04846	0.04947	0.05050	0.05155	0.05262	0.05370	0.05480
–1.5	0.05592	0.05705	0.05821	0.05938	0.06057	0.06178	0.06301	0.06426	0.06552	0.06681
–1.4	0.06811	0.06944	0.07078	0.07215	0.07353	0.07493	0.07636	0.07780	0.07927	0.08076
–1.3	0.08226	0.08379	0.08534	0.08692	0.08851	0.09012	0.09176	0.09342	0.09510	0.09680
–1.2	0.09853	0.10027	0.10204	0.10383	0.10565	0.10749	0.10935	0.11123	0.11314	0.11507
–1.1	0.11702	0.11900	0.12100	0.12302	0.12507	0.12714	0.12924	0.13136	0.13350	0.13567

Z	0.00	0.01	0.02	0.03	0.04	0.05	0.06	0.07	0.08	0.09
–1.0	0.13786	0.14007	0.14231	0.14457	0.14686	0.14917	0.15151	0.15386	0.15625	0.15866
–0.9	0.16109	0.16354	0.16602	0.16853	0.17106	0.17361	0.17619	0.17879	0.18141	0.18406
–0.8	0.18673	0.18943	0.19215	0.19489	0.19766	0.20045	0.20327	0.20611	0.20897	0.21186
–0.7	0.21476	0.21770	0.22065	0.22363	0.22663	0.22965	0.23270	0.23576	0.23885	0.24196
–0.6	0.24510	0.24825	0.25143	0.25463	0.25785	0.26109	0.26435	0.26763	0.27093	0.27425
–0.5	0.27760	0.28096	0.28434	0.28774	0.29116	0.29460	0.29806	0.30153	0.30503	0.30854
–0.4	0.31207	0.31561	0.31918	0.32276	0.32636	0.32997	0.33360	0.33724	0.34090	0.34458
–0.3	0.34827	0.35197	0.35569	0.35942	0.36317	0.36693	0.37070	0.37448	0.37828	0.38209
–0.2	0.38591	0.38974	0.39358	0.39743	0.40129	0.40517	0.40905	0.41294	0.41683	0.42074
–0.1	0.42465	0.42858	0.43251	0.43644	0.44038	0.44433	0.44828	0.45224	0.45620	0.46017
0.0	0.46414	0.46812	0.47210	0.47608	0.48006	0.48405	0.48803	0.49202	0.49601	0.50000

Appendix B: Six Sigma Glossary

2-Bin: A type of pull system using two "bins" (a bin could be a card, or tote, or physical location, etc.). The basic mechanics are as follows. Two bins of material are located at a station. When the first bin is emptied, the operator sends the empty bin to a location to be refilled and begins working from the second bin. The first bin will be returned full prior to the emptying of the second bin.

5S: A process and method for creating and maintaining an organized, clean, and high-performance workplace. The 5S's are Sort, Set in Order, Shine, Standardize, and Sustain.

ABC Analysis: Used to rank order purchased parts and material according to the annual dollar value spent on each. More emphasis should be placed on the few parts that account for most of the cost.

Acceptance Sampling: Inspection of a sample from a lot to decide whether to accept or not accept that lot. There are two types: attributes sampling and variables sampling. In attributes sampling, the presence or absence of a characteristic is noted in each of the units inspected. In variables sampling, the numerical magnitude of a characteristic is measured and recorded for each.

Acceptance Sampling Plan: A specific plan that indicates the sampling sizes and the associated acceptance or nonacceptance criteria to be used. In attributes sampling, for example, there are single, double, multiple, sequential, chain, and skip-lot sampling plans. In variables sampling, there are single, double, and sequential sampling plans.

Acceptable Quality Level: When a continuing series of lots is considered, a quality level that, for the purposes of sampling inspection, is the limit of a satisfactory process average.

Adaptive Control: A defect prevention method that detects errors or possible errors during processes before they can become defects.

Affinity Diagram: An organization of individual pieces of information into groups or broader categories.

Aliasing: When two factors or interaction terms are set at identical levels throughout the entire experiment (i.e., the two columns are 100% correlated).

Alpha Risk: *See* Type I Error.

Alternative Hypothesis: The hypothesis to be accepted if the null hypothesis is rejected. It is denoted by Hl.

Analysis of Means (Anom): A statistical procedure for troubleshooting industrial processes and analyzing the results of experimental designs with factors at fixed levels.

Analysis of Variance (ANOVA): A basic statistical technique for analyzing experimental data. It subdivides the total variation of a data set into meaningful component parts associated with specific sources of variation to test a hypothesis on the parameters of the model or to estimate variance components. There are three models: fixed, random, and mixed.

ANOVA (Analysis of Variance): A statistical test for identifying significant differences between process or system treatments or conditions, performed by comparing the variances around the means of the conditions being compared.

Attribute Data: Go/no-go information. The control charts based on attribute data include percent chart, number of affected units chart, count chart, count-per-unit chart, quality score chart, and demerit chart.

Availability Level: One of the three measures used to calculate overall equipment effectiveness. It represents the percent of time a machine is available to run parts.

Average: The average of a sample (x-bar) is the sum of all the responses divided by the sample size.

Balanced Design: A two-level experimental design is balanced if each factor is run the same number of times at the high and low levels.

Bar Chart: A graphical method depicting data grouped by category.

Batch: A run of like products/parts through a process (number of product/parts run between product changeover).

Benchmarking: An improvement process in which a company measures its performance against that of best-in-class companies, determines how those companies achieved their performance levels, and uses the information to improve its own performance. The subjects that can be benchmarked include strategies, operations, processes, and procedures.

Beta Risk: *See* Type II Error.

Bias: Systematic error that leads to a difference between the average result of a population of measurements and the true accepted value of the quantity being measured.

Bill of Material (BOM): The listing of all components used to make up an assembly. The relationship between the end item assembly (bike) and all lower level items or assemblies (wheel, seat, etc.).

Bimodal Distribution: A frequency distribution that has two peaks. Usually an indication of samples from two processes incorrectly analyzed as a single process.

Binomial Distribution: Given that a trail can have only two possible outcomes (yes/no, pass/fail, heads/tails), of which one outcome has probability p and the other probability $q = 1 - p$, the probability that the outcomes represented by p occurs x times in n trials is given by the binomial distribution.

Black Belt: An individual who receives approximately four weeks of training in the Six Sigma DMAIC methodology, analytical problem solving, and change management methods. A Black Belt is a full-time Six Sigma team leader solving problems under the direction of a Six Sigma Champion.

Block Diagram: A diagram that shows the operation, interrelationships, and interdependencies of components in a system. Boxes, or blocks (hence the name), represent the components; interrelationships could use connecting lines.

Breakthrough Improvement: A rate of improvement at or near 70% over baseline performance of the as-is process characteristic.

C Chart: Count chart.

Capability: A comparison of the required operation width of a process or system to its actual performance width. Expressed as a percentage (yield), a defect rate *(DPM, DPMO)*, an index *(Cp, Cpk*, Pp, *Ppk)*, or as a sigma score (Z).

Cause-and-Effect Diagram: A tool for analyzing process dispersion. It is also referred to as the Ishikawa diagram, because Kaoru Ishikawa developed it, and the fishbone diagram, because the complete diagram resembles a fish skeleton. The diagram illustrates the main causes and subcauses leading to an effect (symptom). The cause-and-effect diagram is one of the seven tools of quality.

Central Composite Design: A three-level design that starts with a two-level fractional factorial and some center points. If needed, axial points can be tested to complete quadratic terms. Used typically for quantitative factors and designed to estimate all linear effects plus desired quadratics and two-way interactions.

Central Limit Theorem: If samples of size n are drawn from a population and the values of x are calculated for each sample, the shape of the distribution is found to approach a normal distribution for sufficiently large n. This theorem allows one to use the assumption of a normal distribution when dealing with x. "Sufficiently large" depends on the population's distribution and what range of x is being considered; for practical purposes, the easiest approach may be to take a number of samples of a desired size and see if their means are normally distributed. If not, the sample size should be increased. This theorem is one of the most important results in all of statistics and is the heart of inferential statistics.

Central Tendency: A measure of the point about which a group of values is clustered; two measures of central tendency are the mean and the median.

Champion: A Six Sigma leader, who recognizes, defines, assigns, and supports the successful completion of Six Sigma projects; a Six Sigma Champion is accountable for the results of projects and the business roadmap to achieve Six Sigma results within their span of control.

Characteristic: A process input or output that can be measured and monitored.

Chi-Square: The test statistic used when testing the null hypothesis of independence in a contingency table or when testing the null hypothesis of a set of data following a prescribed distribution.

Chi-Square Distribution: The distribution of chi-square statistics.

Coefficient of Determination (R^2): The square of the sample correlation coefficient, a measure of the part of variable that can be explained by its linear relationship with a variable; it represents the strength of a model. $(1 - R^2) \times 100\%$ is the percentage of noise in the data not accounted for by the model.

Coefficient of Variation: Defined as the standard deviation divided by the mean (*s I X*bar). It is the relative measure of the variability of a random variable. For example, a standard deviation of 10 microns would be extremely small in the production of bolts with a nominal length of 2 inches but would be extremely high for the variation in line widths on a chip whose mean width is 5 microns.

Common Causes of Variation: Those sources of variability in a process that are truly random, that is, inherent in the process itself.

Complexity: The level of difficulty to build, solve, or understand something based on the number of inputs, interactions, and uncertainties involved.

Confidence Interval: Range within which a parameter of a population (e.g., mean, standard deviation) may be expected to fall, on the basis of a measurement, with some specified confidence level or confidence coefficient.

Confidence Limits: The end points of the interval about the sample statistic that is believed, with a specified confidence coefficient, to include the population parameter.

Conformance: An affirmative indication or judgment that a product or service has met the requirements of a relevant specification, contract, or regulation.

Control Chart: A chart with upper and lower control limits on which values of some statistical measure for a series of samples or subgroups are plotted. The chart frequently shows a central line to help detection of performance over time.

Control Limits: Upper and lower bounds in a control chart that are determined by the process itself. They can be used to detect special or common causes of variation. They are usually set at ±3 standard deviations from the central tendency.

Correlation Coefficient: A measure of the linear relationship between two variables.

Cost of Poor Quality (COPQ): The costs associated with any activity that is not done right the first time. It is the financial qualification of any waste that is not integral to the product or service that your company provides.

Cp: A capability measure defined as the ratio of the specification width to short-term process performance width.

CPk: An adjusted short-term capability index that reduces the capability score in proportion to the offset of the process center from the specification target.

Critical-to-Quality (CTQ): Any characteristic that is critical to the perceived quality of the product, process, or system. *See also* significant *Y*.

Critical *X*: An input to a process or system that exerts a significant influence on any one or all of the key outputs of a process.

Customer: Anyone who uses or consumes the output of a process, whether internal or external to the providing organization or provider.

Cycle Time: The total amount of elapsed time from the time a task, product, or service is started until it is completed.

Dashboards: Term for a series of key measures (e.g., the various gauges on a car dashboard that must be monitored while driving).

Defect: An output of a process that fails to meet a defined specification or requirement, such as time, length, color, finish, quantity, temperature, and so on.

Degrees of Freedom: A parameter in the t, F, and x distributions. It is a measure of the amount of information available for estimating the population variance.

Defective: A unit of product or service that contains *at least one defect.*

Deployment: The planning, launch, training, implementation, and management of a Six Sigma initiative within a company.

Design for Six Sigma (DFSS): The use of Six Sigma thinking, tools, and methods applied to the design of products and services to improve initial release performance, ongoing reliability, and life-cycle cost.

Design of Experiments (DOE): An efficient, structured, and proven approach to investigating a process or system to understand and optimize its performance.

DMAIC: The acronym for the five core phases of the Six Sigma methodology: Define, Measure, Analyze, Improve, and Control; used to solve process and business problems through data and analytical methods.

DPMO (Defects per Million Opportunities): The total number of defects observed divided by the total number of opportunities, expressed in events per million. Sometimes called Defects per Million (DPM).

DPU (Defects per Unit): The total number of defects detected in some number of units divided by the total number of those units.

Efficiency: The ratio of the actual product produced to a standard. Calculated by dividing the standard parts per hour by the actual parts per hour.

Entitlement: The best demonstrated performance for an existing configuration of a process or system. It is an empirical demonstration of the level of improvement that can potentially be reached.

Epsilon (ε): Greek symbol used to represent uncertainty or residual error.

Experimental Design: A formal plan that details the specifics for conducting an experiment, such as which responses, factors, levels, blocks, treatments, and tools are to be used.

F Distribution: Distribution of F-statistics.

F Statistic: A test statistic used to compare the variance from two normal populations.

Factor: An assignable cause which may affect the responses (test results) and of which different versions (levels) are included in the experiment.

Factorial Experiments: Experiments in which all possible treatment combinations formed from two or more factors, each being studied at two or more versions (levels), are examined so that interactions (differential effects) as well as main effects can be estimated.

Failure Mode Effect Analysis (FMEA): A procedure in which each potential failure mode in every subitem of an item is analyzed to determine its effect on other subitems and on the required function of the item.

Finance Representative: An individual who provides an independent evaluation of a Six Sigma project in terms of hard and/or soft savings. They are a project support resource to both Champions and project leaders.

Fishbone Diagram: A pictorial diagram in the shape of a fishbone showing all possible variables that could affect a given process output measure.

Flowchart: A graphic model of the flow of activities, material, and/or information that occurs during a process.

Gauge R&R: The quantitative assessment of how much variation (repeatability and reproducibility) is in a measurement system compared to the total variation of the process or system.

Goodness-of-Fit: Any measure of how well a set of data matches a proposed distribution. Chi-square is the most common measure for frequency distributions. Simple visual inspection of a histogram is a less quantitative, but equally valid, way to determine goodness-of-fit.

Green Belt: An individual who receives approximately two weeks of training in the Six Sigma DMAIC methodology, analytical problem solving, and change management methods. A Green Belt is a part-time Six Sigma practitioner who applies Six Sigma techniques to their local area, performing smaller-scoped projects and providing support to Black Belt projects.

Hidden Factory or Operation: Corrective and non-value-added work applied to produce a unit of output generally not properly recognized as unnecessary and a form of waste of time, resources, materials, and cost.

Histogram: A bar chart that depicts the frequency of occurrence (by the height of the plotted bars) of numerical or measurement categories of data.

Hypothesis Tests, Null: The hypothesis tested in tests of significance is that there is no difference (null) between the population of the sample and specified population (or between the populations associated with each sample). The null hypothesis can never be proved true. It can, however, be shown, with specified risks of error, to be untrue; that is, a difference can be shown to exist between the populations. If it is not disproved, one may surmise that it is true. (It may be that there is insufficient power to prove the existence of a difference rather than that there is no difference; that is, the sample size may be too small. By specifying the minimum difference that one wants to detect and *P*, the risk of failing to detect a difference of this size, the actual sample size required, however, can be determined.)

Hypothesis Tests: A procedure whereby one of two mutually exclusive and exhaustive statements about a population parameter is concluded. Information from a sample is used to infer something about a population from which the sample was drawn.

Implementation Team: A cross-functional executive or management team representing multidisciplinary areas of the company, whose charter is to drive the implementation of Six Sigma by defining, documenting, and leading practices, methods, and operating policies.

Input: A resource consumed, utilized, or added to a process or system. Synonymous with the terms X, characteristic, and input variable.

Ishikawa Diagram: *See* fishbone diagram.

Kaikaku: Radical change.

Kaizen: A continuous improvement vehicle for driving quick hit value by implementing "donow" solutions through waste elimination. A key component of the Toyota Production System.

Kaizen: A Japanese term that means gradual unending improvement by doing little things better and setting and achieving increasingly higher standards. Masaaki Imai made the term famous in his book *Kaizen: The Key to Japan's Competitive Success.*

Kanban: Japanese word for signal. It is used in a pull system to signal when production is to start and can take a number of forms (e.g., cards, boards, lights, bins).

Kurtosis: A measure of the shape of a distribution. A positive value indicates that the distribution has longer tails than the normal distribution (platykurtosis), while a negative value indicates that the distribution has shorter tails (leptokurtosis). For the normal distribution, the kurtosis is 0.

LCL: Lower control limit. For control charts, the limit above which the process subgroup statistics must remain when the process is in control. LCL is typically three standard deviations below the centerline.

Least Squares: A method of curve-fitting that defines the best fit as the one that minimizes the sum of the squared deviations of the data points from the fitted curve.

Little's Law: The fundamental relationship between WIP, cycle time, and throughput.

Logistics: All activities involved in the movement and storage of goods from source to final consumer: Sourcing/Purchasing to Materials Management to Physical Distribution.

Long-Term Variation: The observed variation of an input or output characteristic that has had the opportunity to experience the majority of the variation effects that influence it.

Loss Function: A technique for quantifying loss due to production deviations from target values.

Lot Sampling: Inspection process by which a sample of parts is inspected and based on the outcome the entire lot may be accepted or rejected.

Lot Sizing: MRP function that divides the netted demand into appropriate lot sizes to form jobs. Lot sizes group all demand requirements over a given period defined in MRP.

Lower Control Limit (LCL): For control charts, the limit above which the subgroup statistics must remain for the process to be in control; typically three standard deviations below the central tendency.

Lower Specification Limit (LSL): The lowest value of a characteristic that is acceptable.

Main Effects Plot: A graphic display showing the influence a single factor has on the response when it is changed from one level to another. Often used to represent the "linear effect" associated with a factor.

Manufacturing Cycle Efficiency: A measurement of the percent of value-add time through a process. Calculated by dividing the value-add time by the total cycle time.

Master Black Belt (MBB): An individual who has received additional training beyond Black Belt. The MBB is a technical, go-to expert for technical and project issues in Six Sigma. Master Black Belts are qualified to teach and mentor other Six Sigma Belts and support Champions.

Max Loop (Kanban): Represents the maximum amount of inventory a replenishment pull system will ever encounter. Calculated by summing the amount of WIP, cycle stock, and safety stock within a pull system loop.

Mean Time between Failure (MTBF): The average time that is expected between failures for a product or machine for a defined unit of measure (e.g., operating hours, cycles, miles).

Mean: The average of a set of values. We usually use x to denote a sample mean, whereby we use the Greek letter m to denote a population mean.

Median: For a sample the number that is in the middle when all observations are ranked in magnitude.

Measurement: The act of obtaining knowledge about an event or characteristic through measured quantification or assignment to categories.

Measurement Accuracy: For a repeated measurement, it is a comparison of the average of the measurements compared to some known standard.

Measurement Precision: For a repeated measurement, it is the amount of variation that exists in the measured values.

Measurement Systems Analysis (MSA): The assessment of the accuracy and precision of a method for obtaining measurements. *See also* gauge R&R.

Median: The middle value of a data set when the values are arranged in either ascending or descending order.

Metric: A measure that is considered to be a key indicator of performance. It should be linked to goals or objectives and carefully monitored.

Muda: Waste.

Mura: Abnormality.

Muri: Stress or strain.

Natural Tolerances of a Process: *See* control limits.

Noise: Unexplained variability in the response. Typically, due to variables which are not controlled.

Nominal Group Technique (NGT): A technique, similar to brainstorming, used by teams to generate ideas on a particular subject. Team members are asked to silently come up with as many ideas as possible, writing them down. Each member is then asked to share one idea, which is recorded. After all the ideas are recorded, they are discussed and prioritized by the group.

Nominal: For a product whose size is of concern, the desired mean value for the particular dimension; the target value.

Non-Value-Added (NVA): Any activity performed in producing a product or delivering a service that does not add value, where value is defined as changing the form, fit, or function of the product or service and is something for which the customer is willing to pay.

Normal Distribution: The distribution characterized by the smooth, bell-shaped curve; synonymous with Gaussian distribution.

Objective Statement: A succinct statement of the goals, timing, and expectations of a Six Sigma improvement project.

Opportunities: The number of characteristics, parameters, or features of a product or service that can be classified as acceptable or unacceptable.

Out of Control: A process is out of control if it exhibits variations larger than its control limits or shows a pattern of variation.

Output: A resource, item, or characteristic that is the product of a process or system. *See also* *Y* and CTQ.

***p*-Chart:** For attribute data, a control chart of the proportion of defective units (or fraction defective) in a subgroup. Based on the binomial distribution.

***p*-Value:** The probability of making a Type I error. This value comes from the data itself. It also provides the exact level of significance of a hypothesis test.

Pareto Chart: A graphical technique used to quantify problems so that effort can be expended in fixing the "vital few" causes, as opposed to the "trivial many." The Pareto suggests that most effects come from relatively few causes; that is, 80% of the effects come from 20% of the causes. The Pareto chart is one of the seven tools of quality.

Pareto Principle: The general principle originally proposed by Vilfredo Pareto (1848–1923) that the majority of influence on an outcome is exerted by a minority of input factors.

Poka-Yoke: A transliteration of a Japanese term meaning "to mistake-proof." An engineered method that makes it very difficult or impossible to produce a defective part. Does not require human assistance.

Probability: The likelihood of an event or circumstance occurring.

Problem Statement: A succinct statement of a business situation used to bound and describe the problem that a Six Sigma project is destined to solve.

Process: A set of activities, material, and/or information flow that transforms a set of inputs into outputs for the purpose of producing a product, providing a service, or performing a task.

Process Capability Index: The value of the tolerance specified for the characteristic divided by the process capability. There are several types of process capability indexes, including the widely used Cpk and Cp.

Process Certification: The act of establishing documented evidence that a process will consistently produce its required outcome or meet its required specifications.

Process Characterization: The act of quantitatively understanding a process, including the specific relationship(s) between its outputs and the inputs, and its performance and capability.

Process Cycle Efficiency: A measurement of the percent of value-add time through a process. Calculated by dividing the value-add time by the cycle time.

Process Flow Diagram: *See* flowchart.

Process Member: An individual who performs activities within a process to deliver an output, product, or service to a customer.

Process Owner: The individual who has responsibility for process performance and resources, and who provides support, resources, and functional expertise to Six Sigma projects. The process owner is accountable for implementing Six Sigma solutions in processes.

Productivity: Measurement used to represent the percent of time an operation is performing to a standard. Calculated by dividing the standard hours earned by the actual operating time. Can also be calculated by multiplying the utilization by the efficiency.

Pull System: Process that authorizes production as inventory is consumed. A pull system directly responds to plant changes but must be forced to accommodate customer due dates. The Toyota Production System is an example of a classic pull system.

Push System: Process that schedules production based on demand. A push system directly accommodates customer due dates but must be forced to respond to plant changes. MRP is an example of a classic push system.

Quality Loss Function: A parabolic approximation of the quality loss that occurs when a quality characteristic deviates from its target value. The quality loss function is expressed in monetary units: the cost of deviating from the target increases quadratically the farther the quality characteristic moves from the target. The formula used to compute the quality loss function depends on the type of quality characteristic being used. Genichi Taguchi first introduced the quality loss function in this form.

Quality Function Deployment (QFD): A structured method in which customer requirements are translated into appropriate technical requirements for each stage of product development and production. The QFD process is often referred to as listening to the voice of the customer.

Range: A measure of the variability in a data set; the difference between the largest and smallest values in a data set.

Regression Analysis: A statistical technique for determining the mathematical relation between a measured quantity and the variables upon which it depends; includes simple and multiple linear regression.

Repeatability: The extent to which repeated measurements of a particular object with a particular instrument produce the same value. *See also* gauge R&R.

Reproducibility: The extent to which repeated measurements of a particular object with a particular individual produce the same value. *See also* gauge R&R.

Rework: Activities required to correct defects produced by a process.

Risk Priority Number (RPN): In failure mode effects analysis, the aggregate score of a failure mode including its severity, frequency of occurrence, and ability to be detected.

Rolled Throughput Yield (RTY): The probability of a unit going through all process steps or system characteristics with zero defects.

RUMBA: An acronym for Reasonable, Understandable, Measurable, Believable, and Achievable, used to describe a method for determining the validity of customer requirements.

Run Chart: A basic graphical tool that charts a process characteristic over time recording either individual readings or averages over time.

S Chart: Sample standard deviation chart.

Scatter Plot: A chart in which one variable is plotted against another to observe or determine the relationship, if any, between the two.

Screening Experiment: A type of experiment used to identify the subset of significant factors from among a large group of potential factors.

Seiketsu: Japanese term used in 5S program to describe constant adherence to the first three steps and safety. Often referred to as "standardize."

Seiri: Japanese term used in 5S program to describe segregate and eliminate. Often referred to as "sort."

Seiso: Japanese term used in 5S program to describe the daily cleaning process. Often referred to as "shine."

Seiton: Japanese term used in 5S program to describe arrange and identify.

Seven Tools of Quality: Tools that help organizations understand their processes in order to improve them. The tools are the cause-and-effect diagrams, check sheet, control chart, flowchart, histogram, Pareto chart, and scatter diagram.

Shitsuki: Japanese term used in 5S program to describe the motivation to achieve habitual compliance. Often referred to as "sustain."

Short-Term Variation: The amount of variation observed in a characteristic that has not had the opportunity to experience all the sources of variation from the inputs acting on it.

Sigma Score: A commonly used measure of process capability that represents the number of short-term standard deviations between the center of a process and the closest specification limit. Sometimes referred to as sigma level, or simply Sigma. Also called the Z score.

Six Sigma: A proven and proscriptive set of analytical tools, project control techniques, reporting methods, and management techniques that combine to form breakthrough improvements in problem solving and business performance.

Six Sigma Leader: An individual who leads the implementation of Six Sigma, coordinating all of the necessary activities, and who ensures that optimal results are obtained and keeps everyone informed of progress.

Six Sigma Project: A specifically defined effort that states a business problem in quantifiable terms and with known improvement expectations.

Six Sigma Quality: A term used to generally indicate that a process is well within specifications, that is, that the specification range is ±6 standard deviations. The term is usually associated with Motorola, which named one of its key operational initiatives "Six Sigma Quality."

Skewness: A measure of the symmetry of a distribution. A positive value indicates that the distribution has a greater tendency to tail to the right (positively skewed or skewed to the right), and a negative value indicates a greater tendency of the distribution to tail to the left (negatively skewed or skewed to the left). Skewness is 0 for a normal distribution.

Significant Y: The output of a process that exerts a significant influence on the success of the process or customer satisfaction.

SIPOC (Suppliers-Inputs-Process-Outputs-Customers): A visual representation of a process or system where inputs are represented by input arrows to a box (representing the process or system) and outputs are shown using arrows emanating out of the box.

Special Cause Variation: Those nonrandom causes of variation that can be detected by the use of control charts and good process documentation.

Specification Limits: The bounds of acceptable performance for a characteristic.

Stability: A process with no recognizable pattern of change and no special causes of variation.

Standard Deviation: One of the most common measures of variability in a data set or in a population; the square root of the variance.

Statistical Problem: A problem that is addressed with facts and data analysis methods.

Statistical Process Control (SPC): The use of basic graphical and statistical methods for measuring, analyzing, and controlling the variation of a process for the purpose of continuously improving the process. A process is said to be in a state of statistical control when it exhibits only random variation.

Statistical Solution: A data-driven solution with known confidence/risk levels; as opposed to a qualitative or "seat-of-the-pants" solution.

Supplier: An individual or entity that provides an input to a process in the form of resources or information.

Takt Boards: Visual tool used to communicate required production rates, and track progress to that requirement.

Trend: A gradual, systematic change over time (or some other variable).

TSSW (Thinking the Six Sigma Way): A mental model for improvement that perceives outcomes through a cause-and-effect relationship combined with Six Sigma concepts to solve everyday and business problems.

Two-Level Design: An experiment where all factors are set at one of two levels, denoted as low and high (–1 and +1).

Tukey Test: A statistical test to measure the difference between several means and tell the user which ones are statistically different from the rest.

Type I Error: An incorrect decision to reject something (such as a statistical hypothesis or a lot of products) when it is acceptable.

Type II Error: An incorrect decision to accept something when it is unacceptable.

u Chart: Count per unit chart; a control chart of the average number of defects per part in a subgroup.

UCL: Upper control limit. For control charts, the upper limit below which a process statistic must remain to be in control. Typically, UCL is three standard deviations above the centerline.

Uptime: Net operating time less downtime, setup time, and so on.

Upper Control Limit (UCL): The upper limit below which a process statistic must remain to be in control. Typically, this value is three standard deviations above the central tendency.

Upper Specification Limit (USL): The highest value of a characteristic that is acceptable.

Variability: The property of a characteristic, process, or system to take on different values when it is repeated.

Variable Data: Data where values are continuous, and can be meaningfully measured and subdivided; that is, can have decimal subdivisions.

Variables: Quantities that are subject to change or variability.

Variance: A specifically defined mathematical measure of variability in a data set or population. It is the square of the standard deviation.

Variation: *See* variability.

VOB (Voice of the Business): The representation of the needs of the business and the key stakeholders of the business; usually including profitability, revenue, growth, market share, employee satisfaction, and so on.

VOC (Voice of the Customer): The representation of the expressed and nonexpressed needs, wants, and desires of the recipient of a process output, a product, or a service; usually expressed as specifications, requirements, or expectations.

VOP (Voice of the Process): The performance and capability of a process to achieve both business and customer needs; usually expressed in some form of an efficiency and/or effectiveness metric.

Waste: Material, effort, and time that does not add value in the eyes of key stakeholders (customers, employees, investors).

X: An input characteristic to a process or system. In Six Sigma, it is usually used in the expression $Y = f(X)$, where the output (Y) is a function of the inputs (X).

X-Bar and R Charts: Applies to variable data and is used to create control charts for the average and range of subgroups of data.

X-Bar and S Charts: For variable data, control charts for the average and standard deviation (sigma) of subgroups of data.

X-Bar Chart: Average chart.

Y: An output characteristic of a process. In Six Sigma, it is usually used in the expression $Y = f(X)$, where the output (Y) is a function of the inputs (X).

Yellow Belt: An individual who receives approximately one week of training in Six Sigma problem solving and process optimization methods. Yellow Belts participate in Process Management activities, participate in Green and Black Belt projects, and apply concepts to their work area and their job.

Z score: See *Sigma Score.*

Index

M

W

X

Y

Z